Down to Earth

Down to Earth:
Agriculture and Poverty Reduction in Africa

Luc Christiaensen and Lionel Demery

THE WORLD BANK

©2007 The International Bank for Reconstruction and Development / The World Bank

1818 H Street NW
Washington DC 20433
Telephone: 202-473-1000
Internet: www.worldbank.org
E-mail: feedback@worldbank.org

All rights reserved

1 2 3 4 10 09 08 07

This volume is a product of the staff of the International Bank for Reconstruction and Development / The World Bank. The findings, interpretations, and conclusions expressed in this volume do not necessarily reflect the views of the Executive Directors of The World Bank or the governments they represent.

The World Bank does not guarantee the accuracy of the data included in this work. The boundaries, colors, denominations, and other information shown on any map in this work do not imply any judgement on the part of The World Bank concerning the legal status of any territory or the endorsement or acceptance of such boundaries.

Rights and Permissions

The material in this publication is copyrighted. Copying and/or transmitting portions or all of this work without permission may be a violation of applicable law. The International Bank for Reconstruction and Development / The World Bank encourages dissemination of its work and will normally grant permission to reproduce portions of the work promptly.

For permission to photocopy or reprint any part of this work, please send a request with complete information to the Copyright Clearance Center Inc., 222 Rosewood Drive, Danvers, MA 01923, USA; telephone: 978-750-8400; fax: 978-750-4470; Internet: www.copyright.com.

All other queries on rights and licenses, including subsidiary rights, should be addressed to the Office of the Publisher, The World Bank, 1818 H Street NW, Washington, DC 20433, USA; fax: 202-522-2422; e-mail: pubrights@worldbank.org.

DOI: 10.1596/978-0-8213-6854-1

Library of Congress Cataloging-in-Publication Data

Christiaensen, Luc J.
  Down to earth: agriculture and poverty reduction in Africa / Luc Christiaensen and Lionel Demery.
    p. cm. – (Directions in development)
  Includes bibliographical references and index.
  ISBN-13: 978-0-8213-6854-1
  ISBN-10: 0-8213-6854-0
  ISBN-10: 0-8213-6855-9 (electronic)
  1. Agriculture—Economic aspects—Africa, Sub-Saharan. 2. Green Revolution—Africa, Sub-Saharan. 3. Rural poor—Africa, Sub-Saharan. 4. Poverty—Government policy—Africa, Sub-Saharan. I. Demery, Lionel. II. Title.

HD2117.C477 2006
338.10967—dc22   2006032776

Contents

Preface vii
Acknowledgments ix
Abbreviations xi

Chapter 1   Introduction   1

Chapter 2   Conceptual Framework   9

Chapter 3   Participation of the Poor in Growth   13

Chapter 4   The Growth Potential of Agriculture   47

Chapter 5   Agriculture and Growth in the Rest of the Economy   67

Chapter 6   A Sectoral Decomposition of Poverty Change   73

Chapter 7   Concluding Observations   77

Appendix 1   Objectives of and Data Sources for Case Studies   83

Appendix 2   Data Sources and Constructs Used in Case Studies   85

Appendix 3　Decomposition of Changes in the Poverty
　　　　　　Headcount before and after 1995　　　　　　　　89

Appendix 4　Welfare Effect of Productivity and
　　　　　　Output Price Change　　　　　　　　　　　　　91

Country Case Study Papers　　　　　　　　　　　　　　　　93

References　　　　　　　　　　　　　　　　　　　　　　　95

Index　　　　　　　　　　　　　　　　　　　　　　　　　101

Figures
1.1　Evolution of Cereal Yields and Poverty in East and
　　　South Asia and Sub-Saharan Africa, 1981–2001　　　　　3
2.1　Relative Roles of Agricultural Growth and Nonagricultural
　　　Growth in Reducing Poverty　　　　　　　　　　　　　11

Tables
3.1　Geographical Coverage of Poverty Data　　　　　　　　17
3.2　Decomposition of Changes in the Poverty Headcount　　19
3.3　Decomposition of Changes in the Poverty Gap　　　　　22
3.4　Share and Elasticity Components of the Participation
　　　Effect of Sectoral Growth on Headcount Poverty　　　　28
3.5　Anticipated Effects of an Increase in Agricultural
　　　Productivity or Food Price, Depending on a Household's
　　　Position in the Food and Labor Markets　　　　　　　　34
3.6　Staple Crop Marketing Position in Case Study Countries
　　　by Income/Expenditure Quintiles　　　　　　　　　　　40
3.7　Effect of 20 Percent Decrease in Maize Prices on
　　　Household Welfare by Income Quintile and Market
　　　Position　　　　　　　　　　　　　　　　　　　　　　43
4.1　Agricultural and Nonagricultural Growth Rates by Decade
　　　and Region, 1960–2003　　　　　　　　　　　　　　　48
4.2　Average Agricultural and Nonagricultural GDP, Productivity,
　　　and Population Growth Rates by Region, 1960–2003　　49
4.3　Fertilizer Use, Yields, and Returns in Ethiopia, 1999　　56
6.1　Sectoral Decomposition of Changes in Headcount Poverty　75

Preface

This volume revisits the role of agriculture in poverty reduction. Since the 1950s, this role has been the subject of much debate in the development economics literature. Early development theories viewed agriculture as playing a relatively passive role; it was a sector from which resources could be extracted for industrialization. Nonetheless, some observers believed that a resurgent agriculture could play a dynamic role, fostering development in all sectors. Although agriculture was unlikely to deliver more rapid growth than other sectors in the early stages of development, its links with the rest of the economy could more than compensate. The World Bank's introduction of the Poverty Reduction Strategy Papers has rekindled this debate about the role of agriculture.

Can agriculture deliver more poverty reduction than other sectors? On the one hand, agriculture is likely to grow more slowly than other sectors, but on the other hand, it provides a livelihood for many poor people in the developing world. This volume's analysis, which focuses on Sub-Saharan Africa, reveals that agriculture does deliver more poverty reduction than other sectors, especially in the lower-income countries, because it has strong links with other sectors and because poor people participate more in growth from agriculture than in growth from other sectors.

In reaching this conclusion, the volume assembles a range of empirical material. Its cross-country analysis brings together national accounts data on sectoral growth and household survey data on poverty. To complement the cross-country assessment, the volume goes beyond the data averages by exploring the micro evidence from four low-income countries: Ethiopia, Kenya, Madagascar, and Tanzania. This case study evidence supports the overall conclusion that agriculture can deliver poverty reduction in these African countries. The studies provide an even richer empirical perspective on the links between agricultural productivity growth and poverty reduction. They suggest great potential for accelerating the pace of agricultural productivity growth, given appropriate private and public choices. Candidate interventions include water management, weather-based insurance, agricultural extension services, and rural road construction. Although the benefits of irrigation and road construction are huge, so are the costs, and they require careful cost-benefit analyses.

Acknowledgments

The Norwegian-Finnish Environmentally and Socially Sustainable Development Trust Fund provided financial support for this study.

The authors thank Derek Byerlee, Martin Ravallion, Elisabeth Sadoulet, and Robert Townsend for their invaluable feedback; Leandre Bassole for his support; and Jesper Kuhl for his excellent research assistance.

Abbreviations

CGE	computable general equilibrium
SACCO	savings and credit organization
SIO	semi input-output
SRI	Système de Riz Intensive
TFP	total factor productivity
WTP	willingness to pay

CHAPTER 1

Introduction

Support for agriculture in the development process has fluctuated over the past half century. Dual economy models,[1] popular in the 1950s and 1960s, typically treated agriculture as a backward subsistence sector from which "surplus" resources were to be drawn. They provided theoretical support for development strategies focused on industrialization and led to an urban bias in development planning (Lipton 1977) as well as to fiscal systems that systematically overtaxed agriculture (Krueger, Schiff, and Valdes 1988). At the time some prominent economists (Johnston and Mellor 1961 and Schultz 1964) challenged the predominate view of agriculture, arguing that the sector should play a central role in the development process, especially in its early stages. They emphasized the critical contributions of the agricultural sector to growth in nonagricultural sectors, implying that investments and policy reforms in agriculture might yield faster overall economic growth, even though agriculture itself was likely to grow at a slower pace than other sectors. They also emphasized that adoption of science-based technology could substantially accelerate growth in agriculture, as was later demonstrated by the Green Revolution in Asia.

These insights about the potential indirect and direct growth effects of agriculture refocused attention on agricultural development in the

1970s and 1980s. However, in the 1990s, Sub-Saharan Africa's failure to replicate Asia's Green Revolution and limited improvement in agricultural productivity, despite substantial agricultural policy reforms, led many to question agriculture's potential to reduce poverty. The economic miracle in East Asia further fueled this resurgence of "agro-pessimism," especially concerning Sub-Saharan Africa.

This miracle is usually attributed to export-led growth strategies focused on labor-intensive manufacturing, but the large reduction in poverty observed in Asia went hand in hand with a striking improvement in agricultural performance. At the beginning of 1980, both Asia and Sub-Saharan Africa suffered from mass poverty. In 1981 $1 per day headcount poverty (in terms of purchasing power parity) was estimated at 58, 52, and 42 percent in East Asia, South Asia, and Sub-Saharan Africa, respectively (Ravallion and Chen 2004). By the turn of the century, consumption poverty had fallen to 15 percent in East Asia and to 31 percent in South Asia, while poverty in Sub-Saharan Africa increased slightly to about 46 percent. During the same period, cereal yields in East Asia and South Asia increased by 50 and 68 percent, respectively, but decreased in Sub-Saharan Africa by 15 percent (see figure 1.1).

The persistence of poverty and weak agricultural performance in Africa on the one hand, and the sharp decline in poverty and strong agricultural performance in Asia on the other hand, do not necessarily reflect a causal link. They do suggest the need for policy makers to reconsider the contribution that agriculture might make to overall economic development and poverty reduction. This need has become all the more pressing given the recent focus on poverty reduction in development assistance—beginning with the series of Poverty Reduction Strategy Papers initiated by international financial institutions in 1999.

The debate about the role of agriculture in poverty reduction hinges on two sets of issues. First, although most observers acknowledge that faster economic growth leads to faster poverty reduction, they do not agree on whether investments and policy reforms in agriculture (especially in low-income countries) have a higher payoff in terms of economic growth than investments (and reforms) in other sectors. Second, the extent to which poor people participate in the growth process raises the question of how the poverty-reducing effect of growth in agriculture compares with that of an equivalent growth in other sectors.

Comparing the growth potential of agriculture with that of other sectors is complicated by interactions among the sectors. Distinguishing between a sector's direct growth effect (its rate of growth per se) and

Introduction 3

Figure 1.1. Evolution of Cereal Yields and Poverty in East and South Asia and Sub-Saharan Africa, 1981–2001

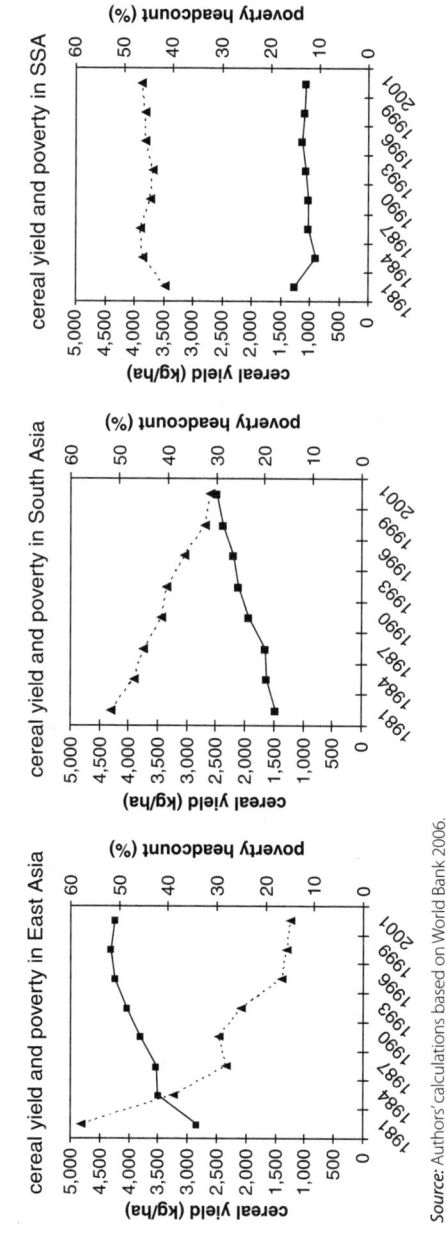

Source: Authors' calculations based on World Bank 2006.
Note: SSA = Sub-Saharan Africa.

indirect effect (the effect of its growth on growth in other sectors) is important. Agriculture tends to grow more slowly than other sectors, but the indirect effects of agriculture on other sectors are larger than the reverse feedback effects. The advantage agriculture holds in terms of its indirect effect might compensate for its disadvantage in terms of its weaker direct effect. Nonetheless, some specialists are concerned about the potential of African agriculture to achieve sufficient (direct) growth to make a dent in poverty, and they doubt whether a Green Revolution can occur in Sub-Saharan Africa. The state of infrastructure, irrigation, human capital, and access to credit in Africa significantly differs from that in Asia when its Green Revolution began. Moreover, the external trading environment is much more competitive today than at that time.

Will the significant indirect growth effects observed in Asia be observed in Africa in the twenty-first century? Delgado, Hopkins, and Kelly (1998) marshal data to show that growing rural incomes will increase demand for domestically produced nonagricultural goods and services, thus potentially enhancing indirect growth effects. But they admit that their estimates depend on a supply response in the nonagricultural sectors—a response that may not be as significant as it has been in Asia. The remoteness of many farming communities, along with a weak and deteriorating infrastructure, will limit the extent to which intersectoral links can flourish and thus adversely affect farmers' ability to buy nonfarm products and urban dwellers' ability to obtain food at lower prices. The neglect of rural infrastructure, a result of entrenched "urban bias" in development strategies (Lipton 1977, 2001), has led to the "hollowing out" of Africa: past investments benefit urban centers in coastal areas, not the rural hinterland (Wood 2002). High transaction costs might therefore undermine the force of the indirect effects of agricultural expansion in Africa, and when transaction costs are lowered, domestic manufacturers may face increasing competition of cheap imports from China.

Agricultural growth generally has a higher return in terms of poverty reduction (that is, a higher "participation effect") than an equal amount of growth in other sectors, because the majority of poor people in the developing world (and especially in Sub-Saharan Africa) directly depend on agriculture for their livelihood. But some observers (such as Maxwell, Urey, and Ashley 2001) are pessimistic about the potential of African smallholder agriculture in the twenty-first century. They point to the complexity of recent technological changes and the stringent quality standards for many food products, both of which are associated with the globalization of commodity chains. In their view, such globalization favors large-scale farmers and agribusiness. Therefore, the extent to which poor people would gain from a pro-agriculture strategy and thus agriculturally driven economic growth is questionable.

The debate about the role of agriculture in poverty reduction in Sub-Saharan Africa calls for an empirical assessment of three sets of questions. First, how can agricultural performance be improved and to what extent does investing in agriculture enhance or harm overall economic growth? Second, is participation by the poor in agricultural growth on average higher than their participation in nonagricultural growth, and if so, what conditions determine their participation rate? Third, if promotion of agriculture tends to increase participation by the poor in growth but leads to slower overall economic growth than the promotion of other sectors, which strategy would tend to have the largest payoff in terms of poverty reduction and under which circumstances?

This volume seeks to find answers to these questions. It draws on national accounts growth data and data on poverty from household surveys to establish empirical regularities in relationships that hold across countries. To shed light on variations among and within countries due to differences in endowments and institutional arrangements, it examines farm household consumption and production data from studies in four countries—Ethiopia, Kenya, Madagascar, and Tanzania (see bibliography for details). These country studies were selected on the basis of availability of integrated data sets, the desire to include countries at different stages of development, and policy makers' interest in having their country participate in this research.

Each country study concerns interactions between agricultural growth (especially productivity growth) and poverty reduction as well as determinants of agricultural performance. Data availability and the policy concerns of the country in question determine the objectives and approach of each study. None of the studies offers a comprehensive examination of the determinants of agricultural growth. They are strategic and selective in the domain they cover, reflecting the desire to inform national policy dialog. Nonetheless, the country-specific insights are relevant to policy dialog in many African countries.

The major contributions of this volume are its synthesis of the potential trade-offs between the lower growth effect and the higher participation effect of agricultural development on poverty reduction, its attention to these relationships within the African context,[2] and its country-specific focus on the empirical strength of different factors in determining agricultural performance.

Chapter 2 presents the conceptual framework for the research presented in subsequent chapters. It shows how the effect of growth in agricultural and nonagricultural sectors on poverty arises from two sources: a participation effect and a growth effect.

Chapter 3 demonstrates that the effect of participation in agricultural growth is significantly larger than that in the other sectors: the poorer the country, the greater the difference in the participation effect. Evidence from the country studies largely confirms this finding derived from cross-country analyses. The extent to which poor people benefit from growth in agriculture depends on the immediate effect of increased agricultural productivity on their income as well as the effect of increased aggregate output on agricultural prices and on demand for unskilled labor. The magnitude of the agricultural income effect depends on the share of total household income derived from agricultural production. The net trading position of poor people in the food market and the tradability of the commodity determine the price effect. The labor market effects (through changes in wages and employment) depend on the structure of the labor market.

Chapter 4 synthesizes the evidence on the direct effects of agricultural growth, again beginning with cross-country data and moving to country-specific evidence. The case studies suggest that the scope for raising agricultural productivity, even within the current production frontier, is considerable. Doing so will require, at a minimum, a location-specific understanding of the constraints to increased productivity; more adequate policies to assist households in coping with weather shocks; and increased adoption of modern inputs, which has been constrained by supply-side factors but also by demand-side factors such as limited profitability, lack of access to credit, and limited capacity to cope with shocks.

Chapter 5 reviews evidence on the indirect effect of agricultural growth. It concludes that the feedback effects from growth in agriculture to growth in the nonagricultural sectors are numerically important and on average at least as large as the reverse effects.

Chapter 6 uses the estimation results from the cross-country analyses to assess how the agricultural and nonagricultural sectors contributed to poverty reduction during the 1980s and 1990s. It confirms that a large part of poverty reduction in the low-income countries in Asia and Latin America is attributable to the good performance of their agricultural sectors and that the slight increase in poverty in Sub-Saharan Africa is largely associated with the dismal performance of its agricultural sector. This analysis confirms what was conjectured at the beginning of this chapter—that the contrasting agricultural performance of Asia and Africa has implications for poverty outcomes. Although agriculture contributed less than other sectors to poverty reduction in the middle-income countries, it contributed substantially more than its size in the overall economy.

Chapter 7 discusses the policy implications of these findings. It concludes that enhancing agricultural productivity is a critical starting point in designing effective poverty reduction strategies, especially in low-income countries. It highlights some of the key requirements for raising agricultural productivity—including improving farming practices, encouraging modern input use (such as fertilizers), connecting farmers to markets and services, and helping farmers deal with shocks.

Notes

1. Dual economy models are archetypical models in which the economy is considered to consist of a modern (usually urban) and a traditional (usually rural) sector.
2. Most (empirical) analyses of this nature have focused on Asia and Latin America; see Ravallion and Datt 1996, 2002; Bravo-Ortega and Lederman 2005.

CHAPTER 2

Conceptual Framework

To frame thinking on the role of agriculture and other sectors in poverty reduction, consider the following identity:

$$\frac{dP_i}{P_i} \equiv \left(\frac{dP_i}{P_i}\frac{Y_i}{dY_i}\right)\frac{dY_i}{Y_i}, \qquad (1)$$

where P_i is any (decomposable) measure of poverty, and Y_i denotes per capita gross domestic product (GDP) in country i. The proportionate change in poverty in a country i is identical to the GDP elasticity of poverty (defined as the proportionate change in poverty divided by the proportionate change in per capita GDP)[1] times the proportionate change in Y_i—that is, GDP growth. The first multiplicative term in equation (1) is labeled the *participation effect*, and the second term is labeled the *growth effect*.

Although economic growth is critical for poverty reduction, not all investments and policy reforms generate an equal amount of growth, and not all growth processes generate an equal amount of poverty reduction (World Bank 2000). In particular, the average growth and participation effects may differ substantially across sectors, as Ravallion and Datt (1996, 2002) have empirically demonstrated for India and Ravallion and

Chen (2007) have demonstrated for China. To accommodate such differences, rewrite equation (1) as a weighted sum of the contributions to poverty reduction of each sector:

$$\frac{dP_i}{P_i} \equiv q\left(\frac{dP_i}{P_i}\frac{Y_{ai}}{dY_{ai}}\right)\frac{dY_{ai}}{Y_{ai}} + (1-q)\left(\frac{dP_i}{P_i}\frac{Y_{ni}}{dY_{ni}}\right)\frac{dY_{ni}}{Y_{ni}}, \quad (2)$$

with a denoting agriculture, n nonagricultural sectors, and q any constant ($0 < q < 1$). A meaningful choice for q is the share of agriculture in total GDP ($Y_{ai}/Y_i = s_{ai}$). It follows that $(1-q) = s_{ni} = Y_{ni}/Y_i$, and equation (2) becomes

$$\frac{dP_i}{P_i} \equiv \left(\frac{dP_i}{P_i}\frac{Y_{ai}}{dY_{ai}}\right)s_{ai}\frac{dY_{ai}}{Y_{ai}} + \left(\frac{dP_i}{P_i}\frac{Y_{ni}}{dY_{ni}}\right)s_{ni}\frac{dY_{ni}}{Y_{ni}}. \quad (3)$$

Using lower cases to represent rates of change for P_i and Y_i yields

$$p_i \equiv \varepsilon_{ai} s_{ai} y_{ai} + \varepsilon_{ni} s_{ni} y_{ni}, \quad (4)$$

where y_{ki} is the growth rate of per capita GDP in sector k, ε_{ki} is the elasticity of total poverty with respect to per capita GDP in sector k, and s_{ki} is the share of sector k in total GDP. When $\varepsilon_n = \varepsilon_a$, equation (4) collapses to equation (1), and the source of growth no longer matters in determining the poverty effect of growth. This property of equation (4) is exploited in developing an empirical test to assess whether the GDP elasticity of poverty differs across sectors in chapter 3. To reflect the fact that growth in the nonagricultural sectors is likely to depend on past growth in agriculture (and that growth in agriculture is likely to depend on past growth in the nonagricultural sectors) and that ε_{ki} and s_{ki} are not constant over time, equation (4) is rewritten in a dynamic framework (with t denoting time and $l = 1,..,t$):

$$p_{it} \equiv \varepsilon_{ait} s_{ait-1} y_{ait}(y_{nit-l}) + \varepsilon_{nit} s_{nit-1} y_{nit}(y_{ait-l}), \quad (5)$$

where $p_{it} = (P_{it} - P_{it-1})/P_{it-1}$ and $y_{kit} = (Y_{kit} - Y_{kit-1})/Y_{kit-1}$. Figure 2.1 provides a schematic presentation of equation (5).

Equation (5) and figure 2.1 show that the impact of a sector (for example, agriculture) on poverty depends on how the pace of growth of the sector compares with that of other sectors. This growth

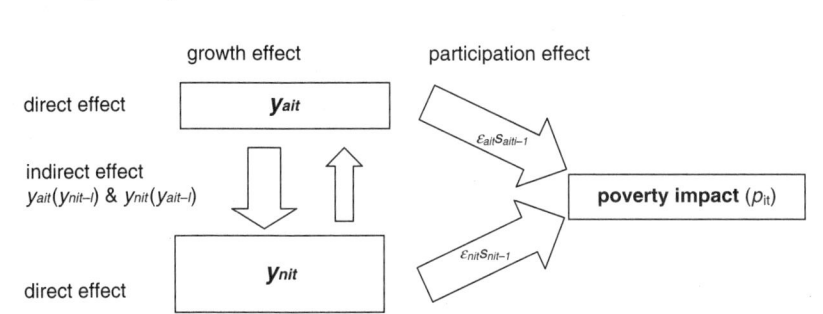

Figure 2.1. Relative Roles of Agricultural Growth and Nonagricultural Growth in Reducing Poverty

Source: Authors.

effect has two components. Its impact on poverty is both direct, flowing immediately from growth in agriculture by raising real incomes of poor farm (and nonfarm) households, and indirect, by accelerating growth in the nonagricultural sectors, $(y_{nit} = y_{nit}(y_{ait-l}))$ (Mellor 1976; Timmer 2005).

The literature highlights three potential interactions between agriculture and overall economic growth. The first is *intersectoral links* forward to agroprocessing activities and backward to input supply sectors. The second is *final demand effects*: given high budget shares spent on food, increases in agricultural output, which lower food prices, would generate meaningful increases in real incomes. These demand effects in turn spur off-farm employment when households choose to buy locally produced nontraded goods and services. The third potential interaction is *wage-goods effects*: agricultural productivity growth would reduce the price of food, thereby reducing real product wages in nonagricultural sectors while raising profitability and investment. Although the reverse interaction might also hold true (nonagricultural growth might spur agricultural growth), the literature suggests that its effects are smaller.

An acceleration in the pace of per capita agricultural growth (y_a) will have a more marked effect on poverty than an identical increase in the rate of nonagricultural growth (y_n) if $\varepsilon_n s_n < \varepsilon_a s_a$.[2] This *participation effect* also has two elements: an elasticity component and a share component. Even though agriculture is a large sector in the economy of most (low-income) developing countries, the share of nonagricultural sectors (services and industry combined) in the overall economy is usually larger than the share of agriculture. Whether the participation effect of agriculture $(\varepsilon_a s_a)$

outweighs the participation effect of other sectors depends, therefore, on whether ε_a is sufficiently larger than ε_n.

In summary, the above-described framework shows that the contribution of a sector to overall poverty reduction depends on its growth and participation effects, each of which has two components. The growth effect has a direct and an indirect component; the participation effect has an elasticity and a share component. All four components have to be taken into account when considering a sector's relative contribution to poverty reduction.[3] The following chapters explore the empirical magnitude of these effects and compare their combined impact on poverty reduction across sectors.

Notes

1. Use of GDP growth rather than the change in mean household income focuses attention on the overall growth process (not simply the growth in household income). The elasticity concept reflects the impact of growth on households' average incomes and the distribution of those incomes. This concept is commonly (but incorrectly) referred to as the "growth elasticity of poverty."
2. Time and country subscripts are dropped to simplify notation.
3. In addition to these four effects, cross-sector population reallocation effects might occur and further contribute to poverty reduction. Anand and Kanbur (1985) and Ravallion and Datt (1996) refer to this process as the Kuznets process. Pursuing this possibility would be empirically challenging because of the difficulty of allocating households across sectors when rural households are often involved in both agricultural and nonagricultural activities. Ravallion and Huppi (1991) provide an alternative decomposition of the proportionate change in poverty that incorporates population shifts between agricultural and nonagricultural sectors.

CHAPTER 3

Participation of the Poor in Growth

This chapter reviews the theory and the empirical literature concerning the participation effect of agricultural growth on poverty. On the basis of a cross-country analysis, it presents new evidence on the difference in participation effects across sectors, with a particular focus on Sub-Saharan Africa (SSA). In addition, the chapter uses evidence from four country studies (Ethiopia, Kenya, Madagascar, and Tanzania) to investigate the factors affecting the magnitude of participation effects.

Theoretical and Empirical Considerations

The contribution of economic growth to poverty reduction might differ across sectors for several reasons. First, capturing the benefits of growth might be easier for poor people if growth occurs where they are located—that is, where the geographical distribution of growth and poverty coincide. Indeed, much of the literature underscoring the importance of agriculture in poverty reduction has argued that poor people stand to benefit much more from an increase in agricultural incomes than from an increase in nonagricultural incomes, because many of the poor live in rural areas, and most of them earn their living in agriculture or agriculture-related activities.[1] This reasoning implicitly assumes that transferring income generated in one economic sector or geographic location to another sector or location is difficult because of market segmentations or considerations of political economy. Second, given that the major asset of poor people is usually their labor, differences in labor intensity across sectors might generate sectoral differences in poverty reduction from growth, as emphasized by Loayza and Raddatz (2005). Third, the distribution of other complementary assets (for example, land in agriculture and capital in industry) may further affect the poverty-reducing effect of growth in different sectors.

The extent to which these phenomena actually result in differences in the poverty-reducing potential of growth across sectors is ultimately an empirical question. Yet few studies have explicitly compared the GDP elasticities of poverty across sectors. Some studies even hint that the GDP elasticities of nonagricultural sectors might be greater than those of agriculture, contrary to what is often implicitly assumed in the above-noted literature. Ravallion and Datt (1996), for example, find that the elasticity of rural headcount poverty with respect to agricultural growth in India is − 0.9, compared with −2.4 for tertiary sector growth. Their conjecture is that the latter is attributable to growth in the informal sector. They find similar orders of magnitude in a more recent paper (Ravallion and Datt 2002).[2] However, for China—and consistent with expectations—Ravallion and Chen (2007) estimate that (aggregate) agricultural growth has the greatest impact on poverty reduction (larger by a factor of four than the impact of aggregate growth in the secondary and tertiary sectors). Findings from these studies underscore the existence of potentially important differences in the sectoral GDP elasticities of poverty across countries, depending on the structure and institutional organization of their economies.[3]

Inspired by this work, other researchers have employed cross-country data (in the absence of sufficiently long country time-series data) to compare GDP elasticities across sectors. Using five-year panel data for the period 1960–2000, Bravo-Ortega and Lederman (2005) find that an increase in agricultural GDP per agricultural worker is not as effective in raising the incomes of the poorest quintile as an equal increase in nonagricultural GDP per nonagricultural worker.[4] Yet, when correcting for the average share of the sector in overall GDP, to obtain sectoral GDP elasticities that can be compared with the results of Ravallion and Chen, growth in agriculture is found to be 2.8 times more effective in reducing poverty than growth in the nonagricultural sectors. Loayza and Raddatz (2005) argue that the labor intensity of the production process is a critical factor in determining the poverty-reducing effect of growth. Linking sectoral growth rates in different countries with data on the intensity of labor use in these sectors and the evolution of poverty, they conclude that growth in agriculture, which is typically the most labor-intensive sector, has the largest potential to reduce poverty, followed by growth in manufacturing, construction, and services. Mining and utilities, which are usually capital-intensive sectors, do not appear to reduce poverty.

What is unclear is whether the more optimistic findings on the agricultural GDP elasticity from China (Ravallion and Chen 2007) and the cross-country evidence in Bravo-Ortega and Lederman (2005) are applicable to the African context. Conditions in Africa are certainly different from those in the wider developing world, including India. Consider initial income inequality, a key determinant of how a given rate of growth reduces poverty. Ravallion (2001) has estimated that for countries with initial Gini coefficients of around 0.60, the GDP elasticity of poverty would typically be –1.2. But if the initial Gini were 0.30, the expected GDP elasticity would be –2.1.

Income or consumption inequalities most likely reflect deeper-seated inequalities. Consider land and education, two assets important for production and income in rural Africa. Economic growth in a rural economy with little landlessness and in which land is relatively equally distributed would be expected to have a greater impact on poverty than growth in a rural economy in which landlessness is pervasive and land is unequally distributed, as for example in India. Bourguignon and Morrisson (1998) find that the larger the share of land cultivated by farmers with small or medium-size operations, the lower the observed income inequality, and thus the larger the effect of growth on poverty. Similarly, the distribution of human capital can profoundly influence the poverty effect of growth. If large sections of the rural population are uneducated and illiterate, they are unlikely to benefit from growth. Farmers who have little access to health services are also less likely to benefit from growth, as are poor African farmers with limited access to other services, such as irrigation, roads, and communications (Christiaensen, Demery, and Paternostro 2005). Because data in sectoral inequalities are not available and data on rural and urban inequality are poor proxies (Bourguignon and Morrisson 1998), the potential influence of inequality differences across the sectors on the participation effect cannot be tested. But these considerations do serve as a reminder that Africa need not follow the pattern found elsewhere in the developing world.

Cross-sectional Variations in the Poverty-Reducing Effect of Growth

Does the poverty-reducing effect of growth vary significantly across sectors? An appropriate empirical specification to answer this question can be derived from equation (5) in chapter 2:

$$\Delta \ln P_{it} = \pi_0 + \pi_{at} s^a_{it-1} \Delta \ln Y^a_{it} + \pi_{mt} s^m_{it-1} \Delta \ln Y^m_{it} + \pi_{vt} s^v_{it-1} \Delta Y^v_{it} + \varepsilon_{it}, \quad (1)$$

where nonagricultural output is disaggregated into output from industry (m) and services (v), and where π_0 and π_{jt} ($j = a, m, v$) are the constant and the sectoral GDP elasticities of poverty to be estimated, and ε_{it} is assumed to be a white-noise error term. It has been shown analytically and empirically that the size of the sectoral GDP elasticity of poverty (π_{jt}) critically depends on the position of the poverty line with respect to the mean of the income distribution, as well as the shape of this distribution (Bourguignon 2003; Klasen and Misselhorn 2006). This position varies substantially, depending on the country's level of development, and thus also evolves over time within countries—an important reason for the time-varying nature of π_{jt}.[5] To account for the effect of the position of the poverty line on the effect of growth on poverty, equation (1) is augmented with interaction terms between the sectoral GDP growth terms and the ratio of the poverty line (z) to average household income/expenditures ($\bar{e}$). This procedure also controls for an important part of the time-varying nature of the elasticities; consequently, π_{jt} is assumed to be constant over time ($\pi_{jt}=\pi_j$):

$$\Delta \ln P_{it} = \pi_0 + [\pi_a + \pi_{ae}(\frac{z}{\bar{e}})]s^a_{it-1}\Delta \ln Y^a_{it} + [\pi_m + \pi_{me}(\frac{z}{\bar{e}})]s^m_{it-1}\Delta \ln Y^m_{it}$$
$$+ [\pi_v + \pi_{ve}(\frac{z}{\bar{e}})]s^v_{it-1}\Delta \ln Y^v_{it} + \varepsilon_{it}. \quad (2)$$

Testing whether the sectoral composition of growth matters for poverty reduction (see Ravallion and Datt 1996; Ravallion and Chen 2007) is now straightforward. When the null hypothesis H_0: $[\pi_a + \pi_{ae}(\frac{z}{\bar{e}})] = [\pi_m + \pi_{me}(\frac{z}{\bar{e}})] = [\pi_v + \pi_{ve}(\frac{z}{\bar{e}})]$ cannot be rejected, equation (2) collapses to a simple regression of the rate of poverty reduction on the rate of growth of GDP (controlling for the position of the poverty line). Under such circumstances, the sectoral composition of growth would not matter, and the debate about the poverty reduction advantages of fostering agriculture rather than nonagricultural sectors becomes the debate about which sector yields the fastest overall economic growth. If, however, the estimated (slope) coefficients in equation (2), controlling for the position of the poverty line, are statistically different, the sectoral composition of growth would also be important for poverty reduction.

To estimate equation (2), poverty measures derived from nationally representative household surveys are brought together with national accounts data on economic growth by sector. The poverty data refer to periods of change or "spells" (derived from two comparable household surveys in years $t - \tau$ and t). Two poverty measures are considered: the headcount index (H) and the poverty gap index (PG). The latter also accounts for the depth of poverty. For comparability, all poverty estimates are based on the $1-a-day poverty benchmark (in purchasing power parity [ppp] terms). The poverty data are from the World Bank's Povcal database (World Bank 2005c).

Table 3.1 provides an overview of these data.[6] It reports 282 spells for 82 countries, 73 percent of the countries having more than one observation.[7] Although the majority of the observations cover periods in the 1990s, about one-third of the spells began in the 1980s (101 of 282). Only five began in 2000. Data on GDP growth for the agricultural and nonagricultural sectors are taken from the World Bank's World Development Indicators and Global Development Finance database (2005a) and are transformed into per capita terms using population statistics from the Food and Agricultural Organization of the United Nations. Per capita sectoral growth rates are in constant 2000 US$, deflated by the overall GDP deflator. In merging sectoral GDP growth with the poverty spells, another 17 spells were dropped due to missing growth data, resulting in a total of 265 spells.

Table 3.1. Geographical Coverage of Poverty Data

Continent	Number of countries	Number of survey periods	Percent of survey periods
Sub-Saharan Africa	20	42	14.9
South Asia	4	22	7.8
East Asia and Pacific	8	47	16.7
East Europe and Central Asia	23	68	24.1
Latin America and the Caribbean	20	89	31.6
Middle East and North Africa	7	14	5.0
Total	**82**	**282**	**100.0**

Source: Authors' calculations.

Estimates were obtained using ordinary least squares regression, and suitable corrections of the standard error were made for heteroscedasticity across countries.[8] The estimates for the poverty headcount are reported in table 3.2. Those for the poverty gap are presented in table 3.3. Cross-country data confirm the expectation that economic growth reduces poverty; the GDP elasticity of headcount poverty is estimated at −3.23 (see table 3.2 panel A). In other words, in the sample countries a 1 percentage point growth in GDP has on average been associated with a 3.23 percent reduction in headcount poverty. The GDP elasticity of PG is −2.95, suggesting that growth also benefits poorer groups.

The results in panel B, table 3.2, highlight the critical importance—ignored in most empirical applications (for example, Ravallion [2001] and Adams [2004])—of controlling for the position of the poverty line in the income distribution in estimating the GDP elasticity of poverty. Inclusion of the poverty line-average income ratio increases the explained variation (R^2) in the observed evolution in poverty from 18 to 21 percent and significantly affects the size of the elasticity. The ratio of the $1 poverty line to the average household income/expenditure was on average 0.33 in the sample. The average for the 25th percentile countries was estimated at 0.14 and for the 75th percentile countries at 0.46, yielding a GDP elasticity of poverty of −4.16 for the former (richer) countries and of −2.1 for the latter (poorer) countries. The increase in the GDP elasticity of poverty as countries get richer should not come as a surprise. This increase is partly driven by numerical logic: an equal percentage point reduction in a variable yields a larger percentage reduction in that variable as the initial value of the variable declines.

The results also show that the composition of growth matters greatly for poverty reduction, as becomes especially clear when the analysis controls for position of the poverty line vis-à-vis the country's income distribution. Although the effect of (aggregate or share weighted) agricultural growth on poverty is larger than that of services and industry, the null hypothesis that $\pi_a = \pi_v$ is rejected, but the null hypothesis that $\pi_a = \pi_m$ is not rejected (p-value equals 0.18; see panel C). However, when the analysis controls for the location of the poverty line (panel D), which increases the explanatory power of the regression, as illustrated by the increase in the test statistic R-squared from 0.23 to 0.27, both the null hypotheses $[\pi_a + \pi_{ae}(\frac{z}{\bar{e}})] = [\pi_v + \pi_{ve}(\frac{z}{\bar{e}})]$ and $[\pi_a + \pi_{ae}(\frac{z}{\bar{e}})] = [\pi_m + \pi_{me}(\frac{z}{\bar{e}})]$ evaluated at the average poverty line-income ratio $[(z/\bar{e}) = 0.33]$ are rejected. At this ratio, a 1 percentage point per capita aggregate (that is,

Table 3.2. Decomposition of Changes in the Poverty Headcount

% change in $1/day headcount poverty	Coeff. A	p-value	Coeff. B	p-value	Coeff. C	p-value	Coeff. D	p-value	Coeff. E	p-value
GDP/cap growth	−3.23	0.000	−5.06	0.000						
GDP/cap growth × (poverty line/average income)			6.41	0.001						
Agricultural GDP/cap growth					−5.81	0.001	−10.03	0.000	−8.93	0.000
Agricultural GDP/cap growth × (poverty line/average income)							9.76	0.000	3.61	0.587
Agricultural GDP/cap growth × (poverty line/average income) × SSA									5.45	0.339
Industrial GDP/cap growth					−3.40	0.000	−4.02	0.000	−3.82	0.002
Industrial GDP/cap growth × (poverty line/average income)							4.38	0.005	2.17	0.728
Industrial GDP/cap growth × (poverty line/average income) × SSA									2.60	0.640
Service GDP/cap growth					−1.08	0.270	−2.02	0.141	−2.19	0.132
Service GDP/cap growth × (poverty line/average income)							3.45	0.084	4.55	0.306
Service GDP/cap growth × (poverty line/average income) × SSA									−0.33	0.928
Constant	0.08	0.124	0.061	0.192	−0.03	0.488	−0.06	0.182	−0.054	0.286
# of observations/R^2	265	0.18	265	0.21	265	0.23	265	0.27	265	0.27

(continued)

Table 3.2. Decomposition of Changes in the Poverty Headcount *(continued)*

% change in $1/day headcount poverty	Coeff. A	p-value	Coeff. B	p-value	Coeff. C	p-value	Coeff. D	p-value	Coeff. E	p-value
test (p-value)										
Agriculture = industry					0.18					
Agriculture = services					0.04					
Agriculture + 0.3 × agriculture = industry + 0.3 × industry							0.005			
Agriculture + 0.7 × agriculture = industry + 0.7 × industry							0.012			
Agriculture + 0.3 × agriculture = service + 0.3 × service							0.001			
Agriculture + 0.7 × agriculture = service + 0.7 × service							0.001			
Agriculture + 0.3 × agriculture + 0.3 × agriculture × SSA = industry + 0.3 × industry + 0.3 × industry × SSA										0.026
Agriculture + 0.7 × agriculture + 0.7 × agriculture × SSA = industry+0.7 × industry + 0.7 × industry × SSA										0.019
Agriculture+0.3 × agriculture+0.3 × agriculture × SSA = service + 0.3 × service + 0.3 × service × SSA										0.001

(continued)

Table 3.2. Decomposition of Changes in the Poverty Headcount *(continued)*

% change in $1/day headcount poverty	Coeff. A	p-value A	Coeff. B	p-value B	Coeff. C	p-value C	Coeff. D	p-value D	Coeff. E	p-value E
Agriculture + 0.7 × agriculture + 0.7 × agriculture × SSA = service + 0.7 × service + 0.7 × service × SSA										0.000
Agriculture = agriculture × (poverty line/ average income) = 0								0.000		
Industry = industry × (poverty line/ average/income) = 0								0.005		
Service = service × (poverty line/average income) = 0								0.221		
Agriculture = agriculture × (poverty line/ average income) = agriculture × (poverty line/average income) × SSA = 0										0.000
Industry = industry × (poverty line/ average income) = industry × (poverty line/average income) × SSA = 0										0.000
Service = service × (poverty line/average income) = service × (poverty line/average income) × SSA = 0										0.240

Source: Authors' calculations based on World Bank 2005a.

Table 3.3. Decomposition of Changes in the Poverty Gap

% change in $1/day poverty gap poverty	Coeff. A	p-value	Coeff. B	p-value	Coeff. C	p-value	Coeff. D	p-value	Coeff. E	p-value
GDP/cap growth	−2.95	0.000	−4.17	0.000						
GDP/cap growth × (poverty line/average income)			4.28	0.027						
Agricultural GDP/cap growth					−4.53	0.008	−7.38	0.002	−6.95	0.005
Agricultural GDP/cap growth × (poverty line/average income)							6.58	0.005	1.79	0.82
Agricultural GDP/cap growth × (poverty line/average income) × SSA									4.88	0.462
Industrial GDP/cap growth					−3.39	0.000	−3.77	0.005	−2.08	0.191
Industrial GDP/cap growth × (poverty line/average income)							2.76	0.140	−9.15	0.264
Industrial GDP/cap growth × (poverty line/average income) × SSA									11.49	0.108
Service GDP/cap growth					−0.98	0.423	−1.31	0.490	−2.43	0.232
Service GDP/cap growth × (poverty line/average income)							1.37	0.619	8.48	0.154
Service GDP/cap growth × (poverty line/average income) × SSA									−5.29	0.28
Constant	0.022	0.693	0.005	0.908	−0.08	0.232	−0.10	0.133	0.087	0.214
# of observations/R^2	265	0.10	265	0.11	265	0.13	265	0.14	265	0.15

(continued)

Table 3.3. Decomposition of Changes in the Poverty Gap *(continued)*

% change in $1/day poverty gap poverty	Coeff. A	p-value A	Coeff. B	p-value B	Coeff. C	p-value C	Coeff. D	p-value D	Coeff. E	p-value E
test (p-value)										
Agriculture = industry						0.57				
Agriculture = services						0.17				
Agriculture + 0.3 × agriculture = industry + 0.3 × industry								0.270		
Agriculture + 0.7 × agriculture = industry + 0.7 × industry								0.420		
Agriculture + 0.3 × agriculture = service + 0.3 × service								0.084		
Agriculture + 0.7 × agriculture = service + 0.7 × service								0.090		
Agriculture + 0.3 × agriculture + 0.3 × agriculture × SSA = industry + 0.3 × industry + 0.3 × industry × SSA										0.135
Agriculture + 0.7 × agriculture + 0.7 × agriculture × SSA = industry + 0.7 × industry + 0.7 × industry × SSA										0.126
Agriculture + 0.3 × agriculture + 0.3 × agriculture × SSA = service + 0.3 × service + 0.3 × service × SSA										0.120

(continued)

Table 3.3. Decomposition of Changes in the Poverty Gap *(continued)*

% change in $1/day poverty gap poverty	Coeff. A	p-value A	Coeff. B	p-value B	Coeff. C	p-value C	Coeff. D	p-value D	Coeff. E	p-value E
Agriculture + 0.7 × agriculture + 0.7 × agriculture × SSA = service + 0.7 × service + 0.7 × service × SSA										0.070
Agriculture = agriculture × (poverty line/average income) = 0								0.008		0.003
Industry = industry × (poverty line/average/income) = 0								0.006		0.001
Service = service × (poverty line/average income) = 0								0.762		0.340
Agriculture = agriculture × (poverty line/average income) = agriculture × (poverty line/average income) × SSA = 0										0.007
Industry = industry × (poverty line/average income) = industry × (poverty line/average income) × SSA = 0										0.003
Service = service × (poverty line/average income) = service × (poverty line/average income) × SSA = 0										0.538

Source: Authors' calculations.

share-weighted) growth in agriculture is 2.65 times more effective in reducing poverty than an equal percentage point aggregate growth in industry. This result is similar to the one derived from the estimation results by Bravo-Ortega and Lederman (2005). Moreover, the ratio of the agricultural over the industrial GDP elasticity of poverty slightly declines from 2.8 to 2.5 for the headcount measure as $z/\bar{e}$ declines from 0.46 (75th percentile) to 0.14 (25th percentile). For the PG measure, it varies between 1.7 and 1.9.

The advantage of agriculture in reducing poverty is even larger when compared with (aggregate) growth in services. The ratio of the elasticities of agricultural growth over service sector growth similarly declines (from 12.8 to 5.6 when using the headcount ratio and from 6.4 to 5.7 when using the PG) when countries become richer ($z/\bar{e}$ decreases from 0.46 to 0.14). The larger ratio of the agricultural over service GDP elasticities of poverty may be related to the imprecise estimation of the effect of growth in services.[9]

When combining industry and services into one sector, the nonagriculture sector (not reported here), the ratio of the elasticities of agricultural growth over nonagricultural growth decline from 4.75 to 3.68 as $z/\bar{e}$ decreases from 0.46 to 0.14. Together these results suggest that the poverty-reducing advantage of growth in agriculture relative to growth in the other sectors becomes smaller as countries become richer. Put differently, the poorer the country, the larger the comparative advantage of agricultural growth in reducing poverty. The constant term is not significantly different from zero (at $p \leq 10$) in the empirical specifications reported in tables 3.2 or 3.3. With zero growth, poverty is predicted to remain unchanged, on average, implying constancy in income/consumption inequality.

In summary, growth generated in agriculture contributes on average more to poverty reduction across the full sample of countries than growth generated in other sectors, irrespective of the poverty measure used. But this advantage of agriculture is likely to decrease as countries become richer. The increase in explanatory power of the regression when decomposing the sources of growth (R^2 increases from 0.21 to 0.27) provides further support for the proposition that the composition of growth matters. An R^2 of 0.27 is very respectable for cross-section type regressions, especially in the absence of country fixed effects, but a substantial amount of variance remains unexplained. This unexplained variance is undoubtedly in part a consequence of the use of data from two sources (household data for the poverty measures and national accounts data for the growth variables) that do not always fully correspond (Ravallion 2003).

Do these results also hold for Sub-Saharan Africa? Only 42 of the 265 observations in the sample are from Sub-Saharan African countries. To estimate the poverty regressions separately for Sub-Saharan Africa would be inappropriate given the small sample. Instead, a Sub-Saharan Africa interaction term is applied to the right-hand-side variables. Panel E of table 3.2 and 3.3 presents the results for H and PG respectively. None of the interactive terms with Sub-Saharan Africa is statistically significant, and the tests reject equality of coefficients in Sub-Saharan Africa for most of the relevant values of ($z/\bar{e}$) (though with slightly less statistical precision for the PG estimates). Therefore, the finding that agricultural growth delivers more poverty reduction than growth elsewhere in the economy appears to hold true for Sub-Saharan Africa. This finding is consistent with the analysis of Dorosh and Haggblade (2003), which is based on a computable general equilibrium model. This analysis reveals that agricultural growth has had a more significant poverty-reducing impact than industry in several African countries, despite the fact that factor productivity growth in manufacturing and agriculture in these countries yielded on average similar overall (direct and indirect) growth effects.

The estimation results presented here were subjected to a series of robustness tests. As indicated above, small percentage *point* changes in poverty could generate large percentage changes. To determine whether such spells might unduly influence the results, the equations above were re-estimated; spells in which the annual change in poverty (that is, the percent change in poverty divided by the length of the spell) exceeded the –100 percent to +100 percent range were excluded. Although 14 observations were dropped, the coefficients remained quasi-unchanged, the R^2 increased to 0.4, and the effect of growth in the service sector became more precisely estimated (p-value of 0.15 as opposed to 0.22 in the full sample). In short, these observations are not driving the results; rather, they have a low signal noise ratio, reducing the precision of the estimation.

A second robustness test involved eliminating strikingly counterintuitive spells of poverty and growth—those with positive (negative) growth greater than 2 percent alongside poverty increases (decreases) of more than 20 percent. The fit of the regression improves substantially despite the loss of 24 observations. The effect of growth in services on poverty now becomes highly significant. Growth in agriculture is estimated to be on average 2.3 times more effective in reducing headcount poverty than growth in non-agriculture (industry and services combined). This is even closer to the results reported by Bravo-Ortega and Lederman (2005).

A third robustness test involves the omission of China—for which Ravallion and Chen (2007) established an (historical) supremacy of agriculture over other sectors in reducing poverty—and India, which together make up more than 10 percent of the sample—20 and 11 spells, respectively. The reported results are robust to the omission of these spells.

Fourth, some observers argue that increasing globalization and liberalization might reduce the advantage agriculture holds over other sectors in reducing poverty. A crude way to examine this hypothesis is to look for structural breaks—that is, a change in the coefficients following a certain time or event. To mark the preglobalization and postglobalization periods, 1995 was chosen, somewhat arbitrarily, as a cutoff point. The results are in appendix 3. The number of observations in the postglobalization (post-1995) subsample drops from 265 in the full sample to 162, and the explanatory power of the specification decreases somewhat. Agriculture continues to hold an advantage over growth in services, but equality of the coefficients for agriculture and industry can no longer be rejected, even though the coefficients are similar in size compared with the full sample. When 1991 is the cutoff point, similar results are obtained (although not reported in the appendix.)

No firm conclusions should be drawn from this partial analysis, but whether globalization affects the advantage of agriculture over the other sectors, especially industry, should be closely watched and studied. On the one hand, globalization potentially exposes smallholder farmers to international market prices, thereby reducing the benefits of domestic agricultural productivity increases on poverty through the price channel. However, given high transaction costs in most rural areas in Sub-Saharan Africa, domestic markets for bulky commodities with high transportation costs, such as food, will likely continue to be isolated from international market prices for quite some time. On the other hand, globalization opens up new markets for nontraditional high-value products such as flowers, fruits, and vegetables. These new markets can reduce poverty through employment generation, especially in and around well-connected regions such as capitals.

Tables 3.2 and 3.3 reflect the response of total poverty to changes in the share-weighted (or aggregate) growth rates of economic sectors. Estimates of the participation effects for each sector can be obtained by multiplying the estimated coefficients by the sectoral shares. Table 3.4 reports the results by region. For the sample as a whole, the participation

Table 3.4. Share and Elasticity Components of the Participation Effect of Sectoral Growth on Headcount Poverty

Region	Number of countries	GDP share (%) Agriculture	GDP share (%) Industry	GDP share (%) Services	Estimated coefficient (=elasticity)[a] Agriculture	Estimated coefficient (=elasticity)[a] Industry	Estimated coefficient (=elasticity)[a] Services	Participation effect of headcount poverty[a] Agriculture	Participation effect of headcount poverty[a] Industry	Participation effect of headcount poverty[a] Services	Ratio of participation effect of Agriculture/ industry	Ratio of participation effect of Agriculture/ services
Lower-income countries												
Sub-Saharan Africa	19	0.32	0.24	0.44	−5.49	−1.98	−0.42	−1.74	−0.48	−0.18	3.63	9.59
South Asia	4	0.29	0.26	0.46	−5.49	−1.98	−0.42	−1.57	−0.51	−0.19	3.09	8.26
East Asia and Pacific	7	0.23	0.41	0.36	−5.49	−1.98	−0.42	−1.24	−0.82	−0.15	1.52	8.28
Eastern and Central Europe	9	0.22	0.33	0.45	−5.49	−1.98	−0.42	−1.22	−0.66	−0.18	1.85	6.61
Latin America and Caribbean	8	0.15	0.28	0.57	−5.49	−1.98	−0.42	−0.82	−0.56	−0.23	1.47	3.51
Middle East and North Africa	1	0.17	0.33	0.51	−5.49	−1.98	−0.42	−0.92	−0.65	−0.21	1.41	4.39
All low-income countries	*48*	*0.25*	*0.31*	*0.44*	*−5.49*	*−1.98*	*−0.42*	*−1.36*	*−0.61*	*−0.18*	*2.23*	*7.41*
Higher-income countries												
Sub-Saharan Africa[b]	2	0.16	0.27	0.56	−8.58	−3.37	−1.51	−1.39	−0.92	−0.85	1.50	1.63
East Asia and Pacific	3	0.14	0.42	0.44	−8.58	−3.37	−1.51	−1.16	−1.42	−0.67	0.82	1.73
Eastern and Central Europe	16	0.16	0.32	0.52	−8.58	−3.37	−1.51	−1.34	−1.08	−0.79	1.24	1.70
Latin America and Caribbean	15	0.11	0.34	0.55	−8.58	−3.37	−1.51	−0.94	−1.14	−0.83	0.82	1.13
Middle East and North Africa	6	0.14	0.32	0.54	−8.58	−3.37	−1.51	−1.24	−1.07	−0.81	1.16	1.53
All middle-income countries	*42*	*0.13*	*0.33*	*0.53*	*−8.58*	*−3.37*	*−1.51*	*−1.12*	*−1.13*	*−0.81*	*1.00*	*1.39*

All countries

Sub-Saharan Africa	20	30	25	45	-6.81	-2.57	-0.88	-2.07	-0.63	-0.40	3.27	5.20
South Asia	4	29	26	46	-6.81	-2.57	-0.88	-1.94	-0.66	-0.40	2.95	4.81
East Asia and Pacific	8	21	41	38	-6.81	-2.57	-0.88	-1.42	-1.06	-0.33	1.34	4.26
Eastern and Central Europe	21	17	32	51	-6.81	-2.57	-0.88	-1.16	-0.83	-0.45	1.40	2.59
Latin America and Caribbean	20	12	32	56	-6.81	-2.57	-0.88	-0.82	-0.84	-0.49	0.98	1.67
Middle East and North Africa	7	15	32	53	-6.81	-2.57	-0.88	-1.01	-0.82	-0.47	1.23	2.13
All countries	**80**	**19**	**32**	**49**	**-6.81**	**-2.57**	**-0.88**	**-1.29**	**-0.83**	**-0.43**	**1.56**	**3.00**

Source: Authors' calculations based on World Bank 2005a.

a. Elasticities were calculated at the average ratio of the poverty line over average income for the pooled sample, and the 75th and the 25th percentile for the lower and higher income group; $(z/\bar{e})$ equals 0.330, 0.465, and 0.149 for pooled countries, lower-income countries, and higher-income countries, respectively. The median ratio $(z/\bar{e})_{50} = 0.24$ was chosen to categorize countries as lower income or higher income. Bold text signifies statistically significant coefficients at $p \leq 0.10$. Other coefficients are not significantly different from zero.

b. The two middle-income countries in Sub-Saharan Africa are Botswana and South Africa.

effect of growth in agriculture is 1.6 times (=−1.29/−0.83) larger than that of growth in the industry sector—that is, one percentage point additional growth in agricultural GDP per capita would reduce the poverty headcount on average 1.6 times more than an additional percentage point growth in industry. The participation effect of growth in agriculture was three times (=−1.29/−0.43) larger than the participation effect of growth in services, though the latter effect was less precisely estimated. These findings suggest that—from the cross-country perspective—fostering agricultural growth is a good starting point for the design of effective poverty reduction strategies.

The results in table 3.4 also underscore the need to look beyond the averages and explore the size of the participation effect across regions and countries. To classify countries as lower income or higher income, the sample was split using the median value of $z/\bar{e}$ (=0.246). The 75th and 25th percentile values of $z/\bar{e}$ were chosen for calculating the elasticity of the lower- and higher-income countries.

As expected, the ratio of the participation effect of agriculture to the participation effect of industry and services is on average larger in the lower-income countries than in the higher-income countries (2.23 versus 1, and 7.41 versus 1.39, respectively). On the basis of the above-noted choice of $z/\bar{e}$, the participation effect of agriculture equals the participation effect of industry in the higher-income countries and is 1.39 times higher than the participation effect of services. This result follows from the much smaller contribution that agriculture makes to overall GDP in the richer countries (13 percent on average, compared with 25 percent in the low-income countries). Despite this smaller contribution, growth in agriculture is more effective in reducing poverty than other sectors, as indicated by the ratio of the share-weighted sectoral GDP elasticities of poverty—that is, once we control for the size of the sector.

Among low-income countries, those in Sub-Saharan Africa obtain the greatest poverty reduction from agricultural growth (the participation effect is −1.74, compared with −0.48 and −0.18 for industry and services, respectively). A 1 percentage point growth in agriculture yields at least 3.6 times as much poverty reduction as a 1 percentage point growth in other sectors in the region.

Overall, these results suggest that fostering agricultural growth is important in reducing poverty in low-income countries, both because of the higher GDP elasticity and the size of agriculture. Such a focus would be especially important in Sub-Saharan Africa, where agriculture is the largest sector of the economy in the poorest countries. The results further indicate that agriculture is also more effective than other sectors in reducing poverty in higher-income countries, though given agriculture's smaller share in the total economy of these countries, a percentage point growth in agriculture may generate only marginally more poverty reduction than an equal percentage point growth in other sectors.

Factors Affecting the Participation Effect from Agricultural Growth

The cross-country analysis shows not only that the participation effect from agricultural growth is on average larger than the participation effect from growth in nonagricultural sectors, but also that the participation effect is likely to differ from country to country. The country studies quantify the participation effect at the country (and even within-country) level and explore the factors affecting the participation effect of an increase in agricultural incomes on household welfare and poverty. In doing so, the studies focus explicitly on the relationships between agricultural incomes and poverty rather than compare the differences in poverty reduction across sectors (the focus of the cross-country analysis discussed above). An increase in agricultural income can be induced through expansion of the basic factors of production (land, labor, capital), an increase in their productivity (that is, better production technologies, use of factor-enhancing inputs such as fertilizer), or a rise in prices received from the output. Land expansion has driven agricultural GDP growth in this volume's case study countries (World Bank 2005d, 2006b). Productivity growth has been limited.[10]

Most of the case studies focus on the effect of enhancing agricultural productivity on household welfare and poverty. However, given that the studies use purposively sampled micro household data, some of the estimates presented below capture mainly first-order effects—that is, the immediate effect of an agricultural productivity increase on household welfare and poverty through its impact on household income. Using data from a community census, the Madagascar case study also traces second-order effects—that is, effects of a (food) productivity increase on poverty through impacts on food prices and wages. In addition, the Kenya case study explores how food price policies per se affect welfare and poverty.

Agricultural Productivity Gains Reduce Poverty in the Case Study Countries

The case studies provide evidence of a significant and direct empirical link between agricultural productivity growth and poverty reduction. The most robust empirical evidence comes from Ethiopia. Dercon and

Christiaensen (2005b) explore the effect of fertilizer use, important in raising agricultural productivity, on welfare and poverty among about 1,500 households in 15 villages across the four major regions of Ethiopia from 1994 to 1999. They find that consumption per adult equivalent among households using fertilizer is on average 8.5 percentage points higher than among households not using fertilizer. They estimate that a doubling of the application rate would on average increase consumption by 2 percent—that is, the elasticity of consumption to fertilizer use (kilogram per hectare) is estimated to be 0.02. Given that the study controls for observable household characteristics (demographic circumstances, rainfall shocks, wealth)—as well as unobserved time-invariant heterogeneity across households (such as household preferences)[11] and time-variant village effects (such as changes in relative prices and the provision of infrastructure)[12]—these results are empirically robust and quantitatively important. Indeed, the effect of fertilizer use on consumption is similar in magnitude to providing an adult female with four years of formal primary education (World Bank 2005a). Were nobody in the sample to use fertilizer, headcount poverty would be, on average, 23 percent, compared with 20 percent if everyone used fertilizer at the current application rate. The poverty gap would decrease from 7.3 to 6.1 percent.

These estimated effects likely represent a lower bound, because they reflect current fertilizer application rates and agronomic practices. Slightly less than half the farmers in the sample use fertilizer, and they apply, on average, only about half the recommended amount (which is about 150 kilograms per hectare[13]). The seemingly low elasticity should be viewed in that light. Doubling application rates among fertilizer users would increase users' consumption, on average, 2 percent. Bringing application rates among those who use no or very little fertilizer (say, 10 kilograms per hectare) up to the current average (about 75 kilograms per hectare) would increase the new users' consumption by 13 percent.[14]

Using cross-section household level data, Sarris, Savastano, and Christiaensen (2006) explore the effect of crop value added per acre on consumption and poverty among a regionally representative sample of rural households in Kilimanjaro and Ruvuma, two cash crop-growing regions in Tanzania. Applying instrumental variable estimation techniques,[15] they estimate the elasticity of consumption per adult equivalent to crop value added per acre at 0.15 in Kilimanjaro and 0.57 in Ruvuma, after controlling for household characteristics and village-level fixed effects. The elasticity in Kilimanjaro is smaller, because the share of agricultural income in total income is smaller in that region than in Ruvuma.

The Tanzania study estimates the effect of agricultural productivity-enhancing interventions on consumption and not on agricultural income only. Relative increases in agricultural income do not translate into equivalent relative increases in consumption. An increase in farm productivity may divert labor from off-farm employment, thereby reducing the overall income and consumption effect (Lopez and Valdes 2000). Simulations suggest that raising the value added per acre for the poor to the median level in the sample would reduce poverty in Kilimanjaro and Ruvuma by 6 and 19 percentage points, respectively. The Tanzania and Ethiopia case studies show that raising agricultural productivity can have substantial and immediate effects on poverty, especially in areas where households derive a significant portion of their income from agriculture (see the next section, Understanding the Links, and table 3.5, for a more formal exposition of this point).

The Kenya case study points to the same conclusion—increases in agricultural productivity having meaningful effects on farmers' incomes. Mistiaen (2006) combined data from experimental farms with survey data from farm households to measure the extent of inefficiency in the production of hybrid maize at different levels of fertilizer use in Kenya. He simulated the first-order impact that such inefficiency would have on poverty by estimating the income losses due to inefficient production (or the income gains to be obtained from reducing inefficiency in maize production). He found that if the price effects were minimal, substantial reductions in poverty could result from decreasing the inefficiency of maize production. On average, achieving optimal productivity could increase the net value of harvested maize per farm household by up to three times the per capita monthly rural poverty figure. Nyoro, Muyanga, and Komo (2005) show that regional trends in poverty among the households in their 1,500 farming household panel in Kenya are closely associated with changes in maize yields. Between 1997 and 2000 and between 2000 and 2004, headcount poverty declined by 25 and 4 percent, respectively, while maize yields increased by 17 and 6 percent. But unlike other studies, the study by Nyoro, Muyanga, and Komo did not control for the influence of other factors through multiple regression techniques. Their results may thus capture both first- and second-order effects, as well as the effects of other factors.

Given the nature of the sample (purposively sampled case studies—Ethiopia and Kenya—or regionally representative studies—Tanzania) and the specification (village fixed effects and no lagged effects of past productivity changes), the estimated coefficients capture only the first-order effects of an increase in agricultural productivity. But the second-order

Table 3.5. Anticipated Effects of an Increase in Agricultural Productivity or Food Price, Depending on a Household's Position in the Food and Labor Markets

Panel A: Monetary value of the change in agricultural productivity: dA>0		Panel B: Monetary value of the change in price: dP>0	
Panel A1: First-order effect; no price or wage effect (dp/dA=dw/dA=0)	$g = p\dfrac{\partial Q}{\partial A}dA = pQ\dfrac{dA}{A}$	**Panel B1: First-order effect; no wage or productivity effect (dw/dp=dA/dp=0)**	$g = [Q-q]dp$
Landless net food buyer ($q>Q=0$; $L^F=0$; $Z=0$)	$g = pQ\dfrac{dA}{A} = 0$	Landless net food buyer ($q>Q=0$; $L^F=0$; $Z=0$)	$g = -qdp<0$
Net food buyer ($q>Q>0$; $L-L^F \geq 0$)	$g = pQ\dfrac{dA}{A} > 0$	Net food buyer ($q>Q>0$; $L-L^F\leq 0$)	$g = [Q-q]dp<0$
Net food seller ($Q>q>0$; $L-L^F\leq 0$)	$g = pQ\dfrac{dA}{A} > 0$	Net food seller ($Q>q>0$; $L-L^F\leq 0$)	$g = [Q-q]dp>0$
Panel A2: First-order and second-order price effect (dp/dA<0; dw/dA=0)	$g = [Q-q]dp + pQ\dfrac{dA}{A}$	**Panel B2: First-order and second-order wage effect (dw/dp>0; dA/dp=0)**	$g = [Q-q]dp + [L-L^F]dw$
Landless net food buyer ($q>Q>0$; $L^F=0$; $Z=0$)	$g = -qdp>0$	Landless net food buyer ($q>Q=0$; $L^F=0$; $Z=0$)	$g > 0$ if $Lw\eta > qp$ or if $\eta > \dfrac{qp}{Lw} = \eta_w^*$
Net food buyer ($q>Q>0$; $L-L^F\geq 0$)	$g = [Q-q]dp + pQ\dfrac{dA}{A} > 0$	Net food buyer ($q>Q>0$; $L-L^F\geq 0$)	$g > 0$ if $[L-L^F]w\eta > [q-Q]p$ or if $\eta > \dfrac{[q-Q]p}{[L-L^F]w} = \eta_w^*$

Net food seller ($Q>q>0$; $L-L^F\leq 0$)		Net food seller ($Q>q>0$; $L-L^F\leq 0$)
$g=0$ if $\frac{Ad(Q-q)}{(Q-q)dA} > -\frac{Adp}{pdA}$		$g>0$ if $[L-L^F]w\eta>[q-Q]p$ or if $\eta<\frac{[q-Q]p}{[L-L^F]w}=\eta_w^*$

Panel A3: First-order and second-order price and wage effect ($dp/dA<0$; dw/dA ?)

Panel B3: First-order and second-order wage and productivity effect ($dw/dp>0$; dA/dp ?)

Landless net food buyer ($q>Q=0$; $L^F=0$; $Z=0$)	$g=[Q-q]dp+pQ\frac{dA}{A}+[L-L^F]dw$	Landless net food buyer ($q>Q=0$; $L^F=0$; $Z=0$)	$g=[Q-q]dp+[L-L^F]dw+p\frac{\partial Q}{\partial A}dA$
	$g>0$ if $Lw\eta<qp$ or if $\eta<\frac{pq}{Lw}=\eta_p^*$		$g>0$ if $wL\eta>pq$ or if $\eta>\frac{pq}{wL}=\eta_{wA}^*$
Net food buyer ($q>Q>0$; $L-L^F\geq 0$)	$g>0$ if $[L-L^F]w\eta<[(q-Q)-\frac{Q}{\varepsilon_{pQ}}]p$	Net food buyer ($q>Q>0$; $L-L^F\geq 0$)	$g>0$ if $[L-L^F]w\eta>[q-(1+\varepsilon_{Ap})Q]p$
	or if $\eta<\frac{[(q-Q)-\frac{Q}{\varepsilon_{pQ}}]p}{[L-L^F]w}=\eta_p^*$		or if $\eta>\frac{[q-(1+\varepsilon_{Ap})Q]p}{[L-L^F]w}=\eta_{wA}^*$
Net food seller ($Q>q>0$; $L-L^F\leq 0$)	$g>0$ if $[L^F-L]w\eta>[(Q-q)+\frac{Q}{\varepsilon_{pQ}}]p$	Net food seller ($Q>q>0$; $L-L^F\leq 0$)	$g>0$ if $[L-L^F]w\eta>[q-(1+\varepsilon_{Ap})Q]p$
	or if $\eta<\frac{[(q-Q)-\frac{Q}{\varepsilon_{pQ}}]p}{[L-L^F]w}=\eta_p^*$		or if $\eta<\frac{[q-(1+\varepsilon_{Ap})Q]p}{[L-L^F]w}=\eta_{wA}^*$

Source: Authors' derivations.

Note: $\eta=\frac{dw}{dp}\frac{p}{w}$, that is, the elasticity of wages with respect to food prices; $\varepsilon_{ij}=\frac{di}{dj}\frac{j}{i}$, that is, the elasticity of i with respect to j.

effects (effects on prices and labor markets) are likely to be quantitatively important, and through the working of the markets, they will affect all households (urban and rural). To illustrate this point, Thirtle and others (2001) estimated the poverty elasticity with respect to crop yields at −0.72 for Africa and at −0.48 for Asia. Datt and Ravallion (1999) estimated the *short-run* elasticity of poverty with respect to agricultural value added per hectare at −0.4 for India. But when they considered indirect effects (through real wage increases), the estimated elasticity was −1.9. However, depending on farmers' net trading position, the second-order effects may erode some or all of the (welfare-enhancing) first-order effects. How changes in the productivity of food production eventually affect the welfare of a population, and the poor population in particular, will depend on several factors, including the distribution of land and other assets, which in turn determines whether a household is a net buyer or net seller of food.

Understanding the Links

Under what conditions do net food buyers/sellers benefit from measures to enhance food productivity or food prices? With respect to productivity, the answer lies in the solution to the utility maximization problem of a farm household that has its preferences defined over consumption (q) and total work effort (L)[16] and that derives its income from household enterprises (that is, farming) and off-farm work. From this equation, the household's welfare gain can be derived from an increase in total (agricultural) factor productivity (TFP) A as given by[17]

$$g \equiv \frac{dv}{\varphi dA} = [Q-q]\frac{dp}{dA} + [L-L^F]\frac{dw}{dA} + p\frac{\partial Q}{\partial A}, \qquad (3)$$

where j is the marginal utility of income, $Q-q$ is the household's net demand for its own produced good (that is, total production Q minus total consumption q), dp/dA is the change in (food) price following an increase in TFP, $L-L^F = L_{nf}-H$ is the household's net engagement in the external labor market (that is, the household's own amount of labor spent outside its own production activities reduced with the amount of labor hired in), dw/dA is the wage change of (agricultural) wage laborers following the increase in TFP, and $p\partial Q/\partial A$ is the monetary value of the change in output (Q) following the increase in TFP (A). Similarly, the household's welfare gain from an increase in the price of the good is given by

$$g \equiv \frac{dv}{\varphi dp} = [Q-q]\frac{dp}{dA} + [L-L^F]\frac{dw}{dp} + \frac{\partial Q}{\partial p}\frac{dA}{dp}. \qquad (4)$$

From equation (3), the effect of a productivity increase on household *welfare* depends on (1) its immediate effect on overall output (first-order effect) as measured by $p\partial Q/\partial A = pQ/A$ (the more the household produces the larger the effect), (2) the extent to which prices and wages are affected by the increase in overall output (second-order effects), (3) whether the household is a net seller ($Q-q>0$) or buyer ($Q-q<0$) of the good, and (4) whether the household is a net seller of labor ($L-L^F>0$) or a net buyer ($L-L^F<0$). The *poverty* effects further depend on (1) whether the poor are producers of the good as smallholders ($pf(.)>0$ for those with land $Z>0$) or as wage laborers ($pf(.)=0$ for those without land $Z=0$), (2) whether the poor are more likely to be net sellers/buyers of the good, and (3) whether the poor are more likely to sell or hire labor. Similar considerations are important when considering the effect of a price increase on household welfare and poverty.

In the discussion that follows, q is assumed to represent a staple crop (for example, rice in Madagascar or maize in Kenya). Three groups capturing most of the poor rural households are distinguished: (1) landless (unskilled) laborers who are typically net food buyers ($Q-q<0$) and who usually earn their income as unskilled agricultural or nonagricultural laborers ($L>L^F=0$); (2) smallholders who are net buyers ($Q-q<0$) and engage in some off-farm employment ($L-L^F>0$); and (3) net sellers of food ($Q-q>0$), usually large-scale farmers who may also hire labor ($L-L^F\leq 0$). Conditions can now be derived under which these groups stand to lose or gain from increases in food productivity, food prices, or both (see table 3.5).

If only first-order effects of an increase in the productivity of staple crop production ($dA>0$; panel A1) are considered, landless farmers are unaffected, while both net buyers and net sellers gain. But when food productivity increases are widespread, they increase aggregate supply and are likely to decrease food prices ($dp/dA<0$; panel A2). The more inelastic the demand for the staple crop, the steeper the price decline. The responsiveness of food prices to internal technical change will further depend on the extent to which local food markets are integrated into international markets. The more isolated the area, region, or country, the higher the transaction costs. In addition, the more isolated the location, the more likely prices in local and national markets will *de facto* be isolated from international markets and thus the more sensitive local prices will be to broadly applied technical change. This finding is discussed below in the context of the Madagascar case study.

All net food buyers (including the landless) stand to benefit from the first-order effect and the effect on food prices of an increase in food productivity, but net food sellers would gain only if the elasticity of *output* with respect to productivity gain exceeds the elasticity of *price* with respect to the productivity gain. Put differently, net sellers would be better off only if output grows faster than prices decrease in response to technical change. Because demand for staple crops is usually inelastic, net sellers often stand to lose and net buyers to gain from technical change. As illustrated below, the majority of net sellers are usually large-scale and rich farmers, whereas many of the net buyers tend to be poor (apart from those engaged in remunerative nonfarm employment). Consequently, in the aggregate, an increase in agricultural productivity is usually associated with a decrease in poverty.

Changes in food prices may in turn affect the wage rates offered to unskilled laborers—often with a lag (Ravallion 1990)—and depending on the responsiveness of wages to food price changes, net food buyers and net food sellers may lose or gain (table 3.5, panel A3). Landless net food buyers will gain from technical change if their loss in wage income (Ldw) is smaller than their gain from reduced food expenditures (qdp). This finding holds true when the elasticity of their wages to food prices (η) is smaller than the ratio of their spending on food (pq) to their income from wage labor (Lw). Similarly, smallholders who are net food buyers stand to gain when their loss in wage income is smaller than their gain from reduced food expenditures. In addition, the more price inelastic the demand for food ($\varepsilon_{Qp}<0$), the lower the gain from their additional food production ($Q/\varepsilon_{pQ}=Q^*\varepsilon_{Qp}$) and thus the more likely that their wage loss exceeds their gain from reduced food expenditures (table 3.5).

Estimating the staple price elasticity of wages (η) is usually constrained by data limitations. But a threshold value η_p^* can be derived from information on a household's farm and off-farm income sources as well as its expenditure patterns from household budget surveys: net buyers and net sellers stand to gain if $\eta<\eta_p^*$ and $\eta>\eta_p^*$, respectively. By way of reference, Boyce and Ravallion (1988) estimated the short-term (instantaneous) elasticity of the wage rate to the price of food grains in Bangladesh at 0.22 and the long-term (steady-state) elasticity at 0.47.

The welfare and poverty effects of a food price increase (for example, through domestic market interventions or tariffs) can be obtained in similar fashion (table 3.5, panel B). Government interventions to increase food prices are often motivated by the interventions' positive effects on the welfare and income of rural households. If only the first-order effect of a food price increase is considered (table 3.5, panel B1), net food buyers stand to lose, while net sellers stand to gain. As a result, the urban poor will lose, and the overall poverty effect will critically depend on the net food position of the rural poor (predominantly net buyers or net sellers). Food price increases may also trigger a change in the labor market and thus an increase in wages ($dw/dp > 0$) (table 3.5, panel B2). Net food buyers would gain if their wage increases sufficiently compensated for their increased food expenses in the wake of food price rises. Conversely, net food sellers stand to gain if their increased wage payments do not erode their increased revenues from higher-priced food. When wage responsiveness to food prices exceeds the ratio of a household's net food spending to its net wage income, net buyers stand to gain and net sellers stand to lose. In addition, food price changes may affect productivity if they provide incentives for investment in staple production. The anticipated welfare and poverty effects are similar to the ones in table 3.5, panel B2, though with a correction of the responsiveness of technical change to food price increases (see table 3.5, panel B3, for exact derivations).

The above framework shows that the impact of agricultural productivity growth on poverty will depend on two main factors: the net marketing position of poor households and the effect of agricultural productivity on food prices and real wages. Similarly, the poverty impact of food price policies, another policy instrument used by governments to increase agricultural incomes, will also heavily depend on the net food marketing position of poor households and on the effect of food price changes on wages and food productivity. The case studies provide evidence on these two key elements.

Poor Households in Sub-Saharan Africa are Typically Net Food Buyers

In the 1980s Weber and others (1988) established—contrary to common assumption—that many poor households in Sub-Saharan Africa are net food buyers. The four country case studies in this volume are based on data from the mid-1990s and early 2000s and confirm that fact (table 3.6). In addition, poor urban households are also net food buyers. As a result, the large majority of poor households stand to gain/lose from (gradual) food price decreases/increases.

Price and Wage Effects are Important in Gauging the Total Poverty Effects of Increased Agricultural Productivity

The findings from the Madagascar case study by Minten and Barrett (2006) corroborate the importance of second-order effects in gauging the impact of a productivity increase. Minten and Barrett use cross-section data

Table 3.6. Staple Crop Marketing Position in Case Study Countries by Income/Expenditure Quintiles

% Rural households Expenditure/ income quintile[a]	Madagascar[b] Net buyers	Madagascar[b] Autarkic	Madagascar[b] Net sellers	Kenya[c] Net buyers	Kenya[c] Autarkic	Kenya[c] Net sellers	Ethiopia[d] Net buyers	Ethiopia[d] Autarkic	Ethiopia[d] Net sellers	Tanzania[e] Net buyer	Tanzania[e] Net sellers
Q1 (bottom)	67	8	25	83	9	8	58	9	34	89	11
Q2	64	14	22	71	14	15	57	6	37	81	19
Q3	67	5	28	69	9	23	56	7	37	72	28
Q4	64	11	25	57	14	28	53	7	40	76	24
Q5 (top)	74	6	20	47	13	40	66	5	29	72	28

Sources: Minten and Barret 2006; Mude 2005; Sarris, Savastano, and Christiaensen 2006; World Bank 2005b.
a. Expenditure quintiles are used for Ethiopia and Tanzania; income quintiles are used for Kenya.
b. Net rice-marketing position of rural population (Minten and Barret 2006).
c. Net maize-marketing position of 1,500 smallholder maize growers in Kenya in 1997 (Mude 2005).
d. Net cereal-marketing position of rural households in Ethiopia in 1995 (World Bank 2005b).
e. Net food-marketing position of rural households in Kilimanjaro (Sarris, Savastano, and Christiaensen 2006).

from a commune census to explain the variance in two nutrition-based poverty variables: the perceived percentage of food-insecure households in the commune and the average length of the households' lean period. They use rice yield as a measure of agricultural productivity. Their tests rejected exogeneity of rice yields, for which they controlled.[18] Their regression specification included a range of conditioning variables such as the geographical and socioeconomic characteristics of the commune. The results suggest that a doubling of rice yields reduces the number of food-insecure households by 38 percentage points and reduces the length of the lean period at the national level by 1.7 months, or about one-third.[19]

Given that Minten and Barrett (2006) use a national census and given the isolated nature of many of the communities, their estimated effects of rice yields on food security measures are likely to reflect both first- and higher-order effects. They explicitly explore this probability and examine how rice productivity influences rice prices and real wages (nominally wage adjusted for rice price), thereby providing an empirical application of the case described in table 3.5, panel A3. By understanding these links, they were able to assess how changes in rice productivity affected three groups of poor households: poor farmers who produce a net surplus of food for the market, poor farmers who are net food buyers, and unskilled workers relying on wage income.

They found that when rice yields double, rice prices fall by only 18 to 45 percent (that is, demand for rice is price elastic), depending on the season, so that farmers with rice surpluses (several of whom are poor) are able to capture a sizeable share (between 10 and 60 percent) of the welfare gains from increasing rice yields. But households with net rice deficits (which represent the bulk of the rural poor) also gain from the induced fall in the rice price.

Finally, Minten and Barret show that a doubling of rice yields would lead to an increase of real agricultural wages of 89 percent during the lean planting season. Households relying on wage income would thus gain from the improvement in rice yields, in part because of lower staple prices, but also because increased labor demand would induce higher nominal wages.

These data from Madagascar clearly show that improvements in agricultural productivity would lead to poverty reduction among all net buyers and also among poor net sellers—at least among those who were not net employers of farm labor ($L-L^F>0$). In other words, the participation effect of productivity-induced agricultural GDP growth will increase as the proportion of net staple crop buyers among the poor increases. The effect further depends on the tradability of the staple crop, which is low in Madagascar due to the country's extremely poor road network. How much poor net buyers gain and whether poor net sellers gain depends on the price elasticity of demand, which also depends on the tradability of the staple crop. In particular, the greater the integration of the production area into the national and international food economy, the more elastic the demand curve faced by producers, the smaller the price decrease, the more likely poor net sellers will gain from productivity increases, and the smaller the benefits accruing to poor net buyers. The participation effect of productivity-induced agricultural growth thus depends on the net food trading position of poor households and the tradability of the staple crop in a given country. Tradability is smaller in Madagascar and Ethiopia than in Kenya, which has better domestically and internationally integrated food markets.

The analysis presented here stops short of a full treatment of the indirect effects of agricultural productivity gains on the nonfarm sector. Although nonfarm workers gain from lower food prices and higher real wages, they would stand to gain even more if growth in the nonagricultural sector increased as a result of increased demand for locally produced goods and services, which enhances employment prospects (and poverty reduction) outside agriculture. Chapter 5 discusses these latter indirect growth effects.

Policies Leading to Higher Food Prices are Likely to Increase Poverty

Governments sometimes seek to increase agricultural incomes through market interventions. Such interventions would likely increase overall poverty if they led to higher food prices. Put differently, an increase in agricultural GDP from policies that increase food prices may reduce the participation effect of the poor in agricultural growth.

Consider Kenya's current policy to help stabilize maize prices. Using a vector autoregression approach, Jayne, Myers, and Nyoro (2005) show that the policy reduced volatility in maize prices, but at the expense of an approximately 20 percent increase in the market price of maize. Mude (2005) demonstrated that the policy (see table 3.5, panel B1) increased the poverty gap by 2.7 percent in urban areas and 2.5 percent in rural areas. As table 3.7 shows, net buying rural households in the lowest quintile are losing the most from the policy.

In essence Kenya's maize price policy transfers a small amount from each household within a large group of net buying urban and rural households,

Table 3.7. Effect of 20 Percent Decrease in Maize Prices on Household Welfare by Income Quintile and Market Position

Income quintile	Percent of households	Net buyers annual per capita income	Relative net gain (%)	Percent of households	Net sellers annual per capita income	Relative net gain (%)	Full sample relative net gain
1	83	2,702	18.13	8	3,135	−3.71	14.82
2	71	7,123	4.44	15	7,896	−1.96	2.85
3	69	13,309	2.19	23	12,675	−1.84	1.09
4	57	21,535	1.30	28	22,774	−1.27	0.38
5	47	54,270	0.51	40	70,910	−0.98	−0.16

Source: Mude 2005.

many of which are poor, to a minority of large maize producers, many of which are rich. Whether welfare gains from reduced maize price volatility under the current policy outweigh the induced welfare loss and increase in poverty is unclear. The productivity gains analyzed by Mistiaen (2006) and reviewed above would appear to have direct benefits not only for maize producers, but also for the majority of poor smallholders, because they are net maize consumers, and the urban poor.

Conclusion

The review of the literature and the new evidence presented here indicate that the participation effect of agricultural growth is substantial—on average, 1.5 to 2 times larger than the participation effect of growth in the nonagricultural sectors. This finding holds true for Sub-Saharan Africa. The case studies demonstrated that the participation effects from agricultural growth depend on the nature of that growth (productivity- or policy-induced food price increases), the net marketing position of the poor (net buyer/seller), and productivity-induced price and wage changes. The case studies showed that many of the poor rural households in Sub-Saharan Africa are net food buyers. As a result, policies that induce higher food prices may moderate the participation effect of the poor in agricultural growth and further impoverish rural and urban populations.

The case studies confirmed that increases in agricultural productivity can lead to important improvements in household welfare and reductions in poverty, especially when secondary (price and wage) effects are taken into account. The extent to which prices change in response to productivity increases depends on the price elasticity of demand and the tradability of the staple crop. The better (staple) markets are integrated, the more tradable the staple crop and the smaller the price decreases and therefore the smaller the benefits to (poor) net buyers and the greater the benefits to (poor) net sellers. However, when markets are poorly integrated, a price collapse may follow a sudden increase in productivity and may induce producers to disinvest the following year (especially when investment concerns reversible technologies such as the use of modern inputs),[20] thereby causing prices to fluctuate and rendering growth in food productivity unsustainable. To effect a gradual price decrease and thus high and sustainable participation of the poor in food productivity-induced agricultural growth, investments in staple crop productivity should be accompanied by investments and policy reforms to foster market integration (World Bank 2005b).

Notes

1. Timmer (2005) and Byerlee, Diao, and Jackson (2005) review this literature.
2. The agricultural output per hectare elasticity of headcount poverty is −0.11, and although the nonagricultural growth elasticities varied significantly across states, they all exceeded the agricultural GDP elasticity.
3. Fan, Chan-Kang, and Mukherjee (2005) report similar findings for China.
4. The regression coefficients of the log of average income of the lowest quintile on the log of agricultural GDP per agricultural worker and the log of nonagricultural GDP per nonagricultural worker are 0.36 and 0.64, respectively, for the low-income countries (excluding Latin America).
5. Other reasons include changes in the endowment structure or the policy environment.
6. Chen and Ravallion (2004) provide a descriptive analysis of the Povcal data.
7. Although observations are not weighted by country population size, China and India, the most populous countries, have the most observations in the sample, and other large countries, such as Brazil and Indonesia, tend to have more than one observation, indicating that the results reported here are implicitly (albeit approximately) population weighted.
8. This approach is comparable to other approaches, for instance Ravallion and Datt (1996) and Adams (2004).
9. More generally, the lack of statistical significance may be due to the difficulties in capturing the informal sector, resulting in measurement error and downward attenuation bias. Government spending often makes up a significant part of GDP in services and, depending on its composition, growth in government spending may not (or not immediately) translate to poverty reduction. Finally, the imprecision in the estimate of service growth coefficient may simply be due to overall noise in the sample. When some of the counterintuitive spells with substantial positive (negative) growth and substantial poverty increase (decrease) are excluded, the explanatory power of the regression increases, and the coefficient on GDP growth in services becomes significant without affecting the other coefficients. But to avoid arbitrary choices in the cleaning procedure, the full sample is used.
10. Annual growth in agricultural value added per agricultural worker between 1990 and 2003 was estimated at 0.5 percent in Ethiopia, −1.5 percent in Kenya, 0.38 percent in Madagascar, and 1.1 percent in Tanzania.
11. Further Hausman-type tests were used to determine whether unobserved time-*variant* household effects bias the estimated coefficients. These tests do not reject exogeneity of fertilizer adoption.
12. To allow for differential effects for richer and poorer households from community-level factors, the time-variant village effects were further interacted

with a dummy variable indicating whether the household had above- or below-median livestock levels.

13. Only 10 percent of all farmers used the recommended amount of 150 kilograms per hectare; half the sampled households used less than 50 kilograms.
14. 6.5*0.02=0.13
15. Instruments for crop value added include rainfall shocks, number of plots cultivated, proportion of land irrigated, and indicator variables for fertilizer and chemical use one year ago.
16. L equals the amount of labor spent on farm (L_f) and nonfarm (L_{nf}) activities.
17. The household is assumed to be a price and wage taker—that is, prices and wages are given for the household. For ease of presentation, consumer and producer prices are also assumed to be equal, though this assumption can be readily relaxed. For more details about the assumptions and the derivation, see appendix 4.
18. The study used the percentage of rice land in the community with improved irrigation infrastructure and the proportion of the commune belonging to a certain ethnic group as instruments.
19. They also find that communities growing cash crops (vanilla and cloves) are more food secure, unlike communities in Tanzania, where the welfare of coffee and cashew growers was not higher (Christiaensen, Hoffmann, and Sarris 2005). But these differences are mostly driven by price fluctuations; at the time of the survey, international prices for cloves and vanilla reached historical highs, and prices for cashew and coffee were in a trough.
20. When the productivity increase follows from irreversible investments such as irrigation or land-conserving infrastructure, producers may find themselves on a price treadmill whereby price decreases completely erode productivity gains and consumers accrue all the benefits.

CHAPTER 4

The Growth Potential of Agriculture

Poverty reduction depends both on the participation of the poor in the growth process and the pace of growth. Chapter 3 indicated that the participation effect of agricultural growth is on average 1.5 to 2 times larger than that of the nonagricultural sectors. But agriculture tends to grow more slowly than nonagricultural sectors. A critical question for policy makers is whether agriculture can grow sufficiently fast (relative to other sectors) to maintain its edge in terms of its participation effect.

This chapter reviews the potential for agricultural growth, beginning with a broad historical cross-country perspective. It also reviews key challenges in enhancing agricultural growth that emerged from the microeconometric productivity analyses in the country case studies.

Potential for Agricultural Growth: A Cross-Country Perspective

A review of the overall sectoral growth rates since 1960 (table 4.1) indicates that agricultural GDP growth has on average lagged behind nonagricultural GDP growth.[1] The difference amounts on average to 1.6 percentage points per year. The gap is largest in South and East Asia and smallest in the Middle East and North Africa. In Sub-Saharan

Table 4.1. Agricultural and Nonagricultural Growth Rates by Decade and Region, 1960–2003

Average annual growth rate (%)	1960–69 Agr.	1960–69 Non-agr.	1970–79 Agr.	1970–79 Non-agr.	1980–89 Agr.	1980–89 Non-agr.	1990–2000 Agr.	1990–2000 Non-agr.	2000–03 Agr.	2000–03 Non-agr.	Total Agr.	Total Non-agr.
Sub-Saharan Africa	2.7	5.0	2.5	5.5	2.6	3.4	2.7	3.0	2.7	3.6	2.6	3.8
South Asia	2.9	5.7	1.7	4.7	3.6	6.4	3.2	6.2	3.0	5.9	2.9	5.8
East Asia and Pacific	4.0	7.7	3.2	7.4	3.0	4.9	1.7	5.1	0.1	5.0	2.3	5.7
East Europe and Central Asia	−1.4	7.0	1.7	7.0	1.3	3.3	−0.7	0.0	3.4	6.7	0.8	2.6
Europe, others	1.2	6.0	1.7	3.5	2.0	2.6	1.7	2.5	−0.8	2.3	1.5	2.9
Latin America and the Caribbean	2.8	5.2	2.3	5.0	1.5	1.6	1.9	3.3	2.2	2.0	2.0	3.3
Middle East and North Africa	1.3	6.1	6.0	7.3	4.8	3.0	3.9	4.2	4.4	3.7	4.4	4.7
North America	—	—	−0.3	3.7	3.2	2.7	2.7	2.7	−1.8	3.2	1.7	3.0
Total	**2.7**	**5.7**	**2.6**	**5.3**	**2.5**	**3.2**	**2.0**	**3.1**	**2.1**	**4.0**	**2.3**	**3.9**

Source: Christiaensen, Demery, and Kuhl 2006.
Note: Both annual agricultural and nonagricultural growth rates are based on GDP expressed in constant 2000 U.S. dollars. Nonagricultural growth is defined as the sector-weighted sum of GDP growth in industry and services.

Africa, the average gap has historically been 1.2 percentage points, somewhat below the world average. The decrease in agricultural growth rates—from 2.7 percentage points in the 1960s to 2.0 percentage points in the 1990s—was accompanied by an even larger decrease in nonagricultural growth rates—from 5.7 percentage points to 3.1 percentage points in the 1990s and 4 percentage points on average during the early 2000s.

The lower growth rates in agriculture have led many policy makers to be skeptical about the potential role of agriculture in development and poverty reduction strategies. Common wisdom holds that agricultural growth rates have not only been historically lower than growth rates outside agriculture, but also that overall productivity growth in agriculture is inferior to overall productivity growth in nonagricultural sectors. This view, not held by some noted authorities,[2] goes back to Adam Smith, who posited that productivity was bound to grow at a slower pace in agriculture than in manufacturing because of agriculture's greater impediments to specialization and labor division. Smith's view is evident throughout the development economics literature[3] and popular among policy makers, but few comparable estimates of productivity growth in agriculture and industry, especially for developing countries, are available.[4] To explore the widely held assumption that improvements in agricultural productivity cannot match those in nonagricultural productivity, sectoral GDP growth rates reported in table 4.1 are divided into their (labor) productivity and population growth components (see table 4.2).[5] The results in table 4.2 suggest that over the past four decades labor productivity in agriculture has on average grown faster than labor productivity in other sectors. The one exception is South Asia. It also appears that overall GDP growth in agriculture has been driven largely by growth in agricultural productivity, while growth in nonagricultural sectors has been fueled largely by population growth in nonagricultural sectors, especially in developing countries (with the exception of those in Eastern Europe and Central Asia). In Sub-Saharan Africa, population growth has been the key contributor to overall growth in agricultural and nonagricultural sectors, but agricultural productivity growth has exceeded nonagricultural productivity growth.

Trends observed in table 4.2 could simply reflect wage-equilibrating movements of labor from agriculture to other sectors in response to higher marginal products of labor (and thus wages) in those sectors. These labor movements would lead to a convergence in sectoral labor productivity and are consistent with the faster (labor) productivity growth in agriculture. Alternatively, the faster (labor) productivity growth in agriculture could result from increased agricultural output due to investment and technological change. If so, productivity increases would free up labor in agriculture and induce workers to move to the nonagricultural sector.

Table 4.2. Average Agricultural and Nonagricultural GDP, Productivity, and Population Growth Rates by Region, 1960–2003

Average annual growth rates (%)	Agricultural GDP	Labor productivity in agriculture	Population in agriculture	Non-agricultural GDP	Labor productivity in nonagricultural sectors	Population in nonagricultural sectors
Sub-Saharan Africa	2.6	0.91	1.7	3.8	−0.64	4.5
South Asia	2.9	1.2	1.6	5.8	2.2	3.6
East Asia and Pacific	2.3	2.9	−0.5	5.7	2.7	2.9
Eastern Europe and Central Asia	0.81	3.4	−2.5	2.6	1.4	1.2
Europe, others	1.5	4.6	−3	2.9	2	0.87
Latin America and the Caribbean	2	2.3	−0.24	3.3	0.48	2.8
Middle East and North Africa	4.4	4.3	0.21	4.7	0.26	4.4
North America	1.7	3.9	−2.1	3	1.8	1.2
Total	**2.3**	**2.4**	**−0.05**	**3.9**	**0.74**	**3.1**

Source: Christiaensen, Demery, and Kuhl 2006.
Note: Figures for the total population in agricultural and nonagricultural sectors were obtained from U.N. Food and Agricultural Organization statistics.

According to this interpretation, the productivity gains in agriculture are the cause of the labor movements (and not its consequence).

Without additional evidence, the relative merits of these hypotheses cannot be ascertained. Nonetheless, while admittedly crude and partial, the descriptive findings in table 4.2 indicate that the notion that agriculture is a backward sector with inherently inferior productivity growth deserves further scrutiny. Both industrial pull and agricultural push are likely to have been at work in recent history.

The limited empirical evidence in the literature, mostly from industrialized countries, would support the hypothesis that total factor productivity (TFP) growth in agriculture does not lag behind TFP growth in other sectors. Using a cost function approach to estimate rates of sectoral TFP growth for the U.S. economy between 1948 and 1979, Jorgenson, Gollop, and Fraumeni (1987, table 6.7) found that TFP growth had been more rapid in agriculture than in almost all other sectors. Using a production function approach to examine historical TFP growth rates of agriculture vis-á-vis the rest of the economy in Australia, Lewis, Martin, and Savage (1988) come to a similar conclusion. Bernard and Jones (1996) estimated average TFP growth at 2.6 percent per year in agriculture, compared with 1.2 percent in industry and 0.7 percent in services, in 14 industrialized countries between the early 1970s and the late 1980s. In only one country was TFP growth higher in industry than in agriculture.

Faster TFP growth in agriculture is also evident in the developing world. Using a production approach applied to panel data for about 50 low- and middle-income countries over the period 1967–2002, Martin and Mitra (2001) found annual TFP growth in agriculture to be on average 0.5 to 1.5 percentage points larger than in nonagricultural sectors, depending on the estimation technique. This difference was statistically significant and valid across the development spectrum.

In summary, the historical evidence reviewed here questions the view of agriculture as a backward sector with limited inherent growth potential and thus a limited direct growth effect on poverty reduction. Although agriculture has been growing more slowly than other sectors, agricultural productivity appears to grow at least as fast as nonagricultural productivity, and studies comparing productivity growth suggest that the faster agricultural productivity growth does not primarily follow from equilibrating labor movements, but rather from a more rapid increase in total factor productivity in agriculture per se. Attempts to increase agricultural productivity as a key building block of poverty reduction strategies should therefore not be rejected offhand.

However, the agricultural sector should not be expected to grow faster than the nonagricultural sector or to increase its share in the economy. Engel's Law implies that as incomes rise, demand for agricultural products increases at a slower rate than demand for nonagricultural products, and hence the share of agriculture in total output declines. This law is consistent with the historical pattern of migration between agricultural and nonagricultural sectors observed in the data. In other words, although the direct growth effect of agriculture on poverty reduction will likely continue to be smaller than that of other sectors, experience shows that agricultural productivity and growth can be increased substantially, a phenomenon necessary to facilitate migration of labor out of agriculture (Gollin, Parente, and Rogerson 2002) and to foster nonagricultural growth (Irz and Roe 2005).

Enhancing Agricultural Productivity: Selected Policy Themes

Despite faster productivity growth in agriculture compared with other sectors, agricultural (labor) productivity growth historically has been low in Sub-Saharan Africa. How could it be increased and at what cost? Answers are likely to be highly context specific, and experience suggests that a package of well-targeted interventions reflecting understanding of local constraints is necessary.

A detailed exploration of ways to increase agricultural productivity in Sub-Saharan Africa is beyond the scope of this volume. Although the country case studies explore the determinants of productivity in general terms, they focus on themes of particular policy interest in each country. Nonetheless, some of the study findings are likely to be relevant to many countries in Sub-Saharan Africa. That the findings pertain to technology-related issues more so than to institutional and marketing arrangements should not be viewed as a judgment on those arrangements' relative importance in improving agricultural productivity. Policy messages from the findings of the case studies are highlighted below.

Policy message 1: Scope for increasing agricultural growth appears substantial, because staple crop farmers' production is well within the production frontier and because field experiments suggest the potential for expanding this frontier is great.

Cereal yields in Madagascar are estimated at 2,380 kilograms per hectare and at between 1,200 and 1,500 kilograms per hectare in Ethiopia, Kenya, and Tanzania—well below the average in South Asia (2,500 kilograms per hectare in 2005) and East Asia and the Pacific (4,518 kilograms per hectare in 2005) (World Bank 2005a). This contrast in yields between case study countries and Asian countries would suggest that substantial gains in the former countries are possible.

Using stochastic frontier analysis to calculate coefficients of technical efficiency, Sarris, Savastano, and Christiaensen (2006) find that agricultural households in Kilimanjaro produce on average about 40 percent less than the region's best performers, who constitute the (relative) production frontier, and farmers in the Ruvuma region produce on average 35 percent less than their best-performing counterparts. These inefficiency estimates reflect controls for land, labor, and input use as well as actual rainfall, soil quality,[6] and land-improving infrastructure such as irrigation and soil erosion protection. They suggest that technical inefficiency in production is substantial. In Ethiopia, average technical efficiency was estimated at about 58 percent—that is, farmers' crop yields were on average 42 percent below the production frontier (World Bank 2006a).

Controlling for seed type, Mistiaen (2006) compares actual maize yields from trial stations in Kenya using different fertilizer combinations with yields obtained by farmers on sites with characteristics similar to the trial stations. He estimates the technical inefficiency of the farmers at 60 percent, suggesting that revised agronomic practices can lead to substantial improvements. Randrianarisoa and Minten (2006) offer supporting evidence of the potential efficacy of such practices in Madagascar. They discovered that halving the planting age at transplantation—a proxy for the adoption of the Système de Riz Intensive (SRI) cultivation technique[7]—is associated with a 13 percent increase in rice yields. Minten (2006) confirmed that substantial improvements are also possible with increased fertilizer use. In Madagascar he found an average return of 7.6 kilograms of paddy rice per kilogram of fertilizer use and estimated that current fertilizer use (by only 6 percent of the country's farmers) would have to increase 25-fold, holding all else constant, to generate rice yields comparable to those obtained in Vietnam.

In addition to technical inefficiency, Sarris, Savastano, and Christiaensen (2006) find evidence of allocative inefficiency. They compare the value of the marginal product of the different factors (land, labor, inputs, capital) with the observed factor price in Ruvuma and

Kilimanjaro and discover that labor is used in excess, whereas inputs and capital (farm equipment) are largely underutilized.

Substantial improvements in agricultural production appear further possible by *expanding* the production frontier. Average cereal yields on experimental plots are much higher than those obtained by farmers. For example, maize yields on experimental plots or farmers' demonstration plots in Ethiopia are three to four times higher than those obtained through traditional practices. Yields for other cereal crops could be more than doubled with appropriate input use and improved crop management (World Bank 2005b). Research systems in Tanzania have produced innovations related to pest and crop disease management, soil fertility, and postharvest losses that could vastly increase crop productivity (World Bank 2005d).

Despite these positive developments, adoption of productivity-enhancing technologies remains limited. At one level this situation reflects a lack of funding devoted to agricultural research and its dissemination, especially in the area of crop production. Minten (2006) reports that current spending on research to improve the productivity of rice production in Madagascar is about 0.1 percent of the total annual value of rice production. Even where the focus on agricultural extension is strong, such as in Ethiopia, improvements have been slow in coming. One reason is low profitability (a too low output/input price ratio) due to limited market access or irregular supply of inputs, limited access to working capital, and farmers' inability to cope with shocks ex post.

The efficiency gap in the agricultural production of Sub-Saharan Africa and that of other regions has been observed for decades (see Ali and Byerlee 1991 for a review of the literature). Evidence from this volume's case study countries highlights this scope once again and is consistent with more recent evidence from other Sub-Saharan African countries.[8] The substantial scope to improve agricultural growth by moving farmers closer to the existing production frontier, which remains well below what experiments show is technically feasible, deserves policy makers' immediate attention.

Policy message 2: The variety of agro-ecological and institutional environments across and within countries underscores the need for location-specific analyses, as well as decentralized research and development and extension systems.

In comparing allocative inefficiency in the agricultural production of Ruvuma and that of Kilimanjaro, Sarris, Savastano, and Christiaensen

(2006) found that land is underutilized (value of marginal product of land > p_{land}) in Kilimanjaro, where the poor are more land-constrained than the rich, but that land is slightly overutilized in Ruvuma. This finding reflects the scarcity of land in Kilimanjaro and the abundance of land in large parts of Ruvuma. Sarris, Savastano, and Christiaensen also find that land-saving inputs are underutilized in both Kilimanjaro and Ruvuma, but much more so in Kilimanjaro—land scarcity increases the opportunity cost of not using inputs. Capital (farm equipment), on the other hand, appears to be more underutilized in Ruvuma than in Kilimanjaro. Further comparison of the relative factor prices (price of labor/price of land) indicates that labor is relatively less expensive than land in Kilimanjaro, compared with Ruvuma. This finding suggests that land-saving technologies, such as fertilizer, improved seeds, and pesticides, might be more appropriate in Kilimanjaro and that labor-saving technologies, such as animal traction and tractors, might be more appropriate in Ruvuma. As a result, institutional arrangements to provide credit to finance the purchase of these goods should probably differ as well.

The Madagascar case study also describes varying demand for labor- versus land-saving technology across different regions. National surveys indicate that access to agricultural equipment and cattle, both of which increase labor productivity, are the most important means to increase agricultural productivity. Surveys from rice growers in the highlands, on the other hand, suggest that technologies to boost land productivity, such as access to agricultural inputs and to cattle for manure, are more critical.

The need for a geographically differentiated rural development strategy has been underscored in Ethiopia (Ehui and Pender 2005). Until recently, the focus was on boosting cereal production across the country, despite varying agro-ecological characteristics. The somewhat disappointing result has been partly ascribed to the undifferentiated top-down approach to fostering adoption of modern inputs (World Bank 2006a). Now the government has adopted a strategy with location-specific recommendations concerning optimal product mix (cereals, vegetables, livestock, and forestry) as well as optimal input package (fertilizer types, pesticides, and improved seeds) (World Bank 2005b).

The need for a differentiated approach emerging from the findings of this volume's case studies is consistent with the emphasis of other studies on different factors in explaining farm inefficiency in Sub-Saharan Africa (Ajibefun, Battese, and Daramola 2002; Binam and others 2004;

Kaliba 2004). Despite the variety of conditions among and within countries, four findings were common to the country case studies:

- Important losses are associated with weather-related shocks and poor farmers' limited ex-post ability to cope with shocks through adoption of low-risk, low-return production technologies.
- Profitability, credit access, and risk-coping capacity are important demand-side constraints to adoption of modern inputs.
- Improvements in rural road infrastructure could have big payoffs but are quite costly and therefore should be based on cost-benefit analyses.
- Poorer farmers may face different constraints to increasing agricultural output than richer farmers.

Policy message 3: Better water management and strengthened ex-post coping strategies will be critical in raising agricultural productivity. Both will mitigate the immediate losses from weather-related shocks and will induce poor farmers to adopt high-return production techniques, which are usually more risky than low-production techniques.

Crop income and household welfare losses associated with weather shocks are substantial, even outside arid and semi-arid areas. Sarris, Savastano, and Christiaensen (2006) estimated that gross revenue from crop production was 36 percent lower among households in Kilimanjaro that reported below-normal rainfall on their plots and 59 percent lower among households that reported much-below-normal rainfall. Given a long run average of 1,160 millimeters rainfall per year, Kilimanjaro is usually not thought of as a dry region. Nonetheless, households in Kilimanjaro speak of a drought (rainfall much below normal) when rain is less than 24 percent below the long-term median (about 884 millimeters), and droughts are estimated to occur once every three to four years. Rainfall in Ruvuma is on average similar but more reliable; droughts are reported to occur much less frequently, again underscoring the need for spatially differentiated approaches to increasing agricultural productivity. The strong negative effect of droughts and other weather-related events (for example, cyclones) on crop production and human welfare have been well documented by Dercon and Christiaensen (2005a) and the World Bank (2005b) for Ethiopia and by Minten (2006) for Madagascar.

In addition to the immediate production and welfare losses associated with droughts and other weather-related events, Dercon and Christiaensen (2005a) demonstrate that farmers' inability to cope with shocks ex post affects their ex-ante behavior. In particular, it induces them to adopt low-risk, low-return technologies, thereby forgoing available technologies such as fertilizer, which yield higher returns, but are usually also riskier. Table 4.3 shows that fertilizer use in Ethiopia tends to increase both crop revenue and risks. Although yields in normal years (median rainfall and climatic conditions) and bad years (when the level of rainfall is equal to the 20th percentile in historical rainfall distributions) are systematically higher when fertilizers are used, crop returns are systematically higher in bad years in which no fertilizers are used. Once the cost of credit is added, the difference is even larger. Net crop revenues are only marginally higher when fertilizer is used in normal years because of the low cereal/fertilizer price ratio or poor agronomic practices.

Low profitability of fertilizer use due to a low cereal/fertilizer price ratio is a potentially important factor in explaining low fertilizer use (World Bank 2006a, 2006b). In their estimation of the effect of limited ex-post coping capacity on fertilizer use, Dercon and Christiaensen control for this low profitability through time-varying village-fixed effects. Controlling for poor agronomic practices through household-level fixed effects, they estimated that insuring ex-post consumption so that it would never fall below the *household-specific* median would increase fertilizer application among current fertilizer users by 8 percent. An increase in the level of consumption when the harvest fails by one standard deviation of the cross-sectional distribution of consumption

Table 4.3. Fertilizer Use, Yields, and Returns in Ethiopia, 1999

	All cereals	Teff	Barley	Wheat	Maize	Sorghum
Yields in kilograms per hectare						
Normal year, no fertilizer	419	311	402	451	494	519
Bad year, no fertilizer	307	224	295	313	380	338
Normal year, fertilizer	602	447	579	648	711	747
Bad year, fertilizer	442	322	424	450	547	486
Returns in Birr per hectare						
Normal year, no fertilizer	837	622	804	901	988	1,038
Bad year, no fertilizer	615	447	590	625	760	676
Normal year, fertilizer	847	512	770	928	1,058	1,144
Bad year, fertilizer	527	262	461	531	730	623

Source: Dercon and Christiaensen 2005a.
Note: Yields are estimated using a production function and data from the 1999 Ethiopian Rural Household Survey. The return is the 1999 value of output minus any applicable fertilizer costs (ignoring credit costs).

would induce a 44 percent increase in fertilizer use among those already applying it. An equivalent increase by one standard deviation of livestock holdings, the main portion of liquid assets in this context, would increase fertilizer application rates by a lower percentage (23 percent), suggesting that risk is an important constraint, and possibly more important than working capital constraints.

An important risk-mitigating intervention is better water management (water harvesting/irrigation). Irrigated land remains a small proportion of all cultivated land in the case study countries. In Ethiopia, use of irrigation techniques was only reported on 1.25 percent of all cropped land in 2003/4. Nonetheless, using nationally representative plot-level data, the World Bank (2006a) reports substantial returns to irrigation, especially in Ethiopia's more food-insecure areas. In Kilimanjaro, about 20 percent of the cultivated land is irrigated, yielding a substantial payoff in terms of crop revenue, compared with only 4 percent in Ruvuma.

Although irrigation is effective in raising average crop revenue and household welfare, it appears ineffective in mitigating the effects of widespread drought in Kilimanjaro. The reason is that irrigation in Kilimanjaro is mostly gravitation irrigation. Consequently, when droughts are widespread, the rivers dry up, leaving little or no water for irrigation (Christiaensen, Hoffmann, and Sarris 2005). Nonetheless, the availability of irrigation infrastructure facilitates access to credit, which is critical to adoption of inputs. In addition, access to irrigation enhances adoption of fertilizer in Madagascar (Minten 2006).

The benefits of irrigation infrastructure are substantial, but so are the costs. The cost of rehabilitating the existing irrigated perimeters in Madagascar was estimated at $300 million ($1,000 per hectare) or three times the cost of the ongoing World Bank Rural Development Project (Minten 2006). Careful cost-benefit analysis that accounts for all costs and benefits (including second-order effects through lower crop prices) is needed. In areas not suitable for irrigation, interventions such as drought-resistant seeds and weather-based insurance could strengthen poor farmers' capacity to cope with shocks.

Sarris, Karfakis, and Christiaensen (2006) found substantial stated demand for weather-based insurance schemes in Tanzania. Were the premium to be set at the average willingness to pay (WTP), about one-quarter of households in Kilimanjaro would insure about 85,000 to 120,000 acres (about 40 to 50 percent of total land cultivated), resulting in a consumer surplus or benefit to society of about $1 million, depending on the particular contract. This benefit underscores the welfare loss associated with uninsured risks. About half of all households in Kilimanjaro and about one-third of all households in Ruvuma indicated an interest in weather-based insurance. Liquidity constraints were the main reason for lack of interest.

However, the average WTP in Kilimanjaro constituted only about 30 to 55 percent of the actuarially fair value of the contract, depending on the contract. Given the large consumer surplus of about 4.2 to 11 percent of total crop sales, and given that actuarially fair premiums are in the range of 10 to 20 percent of crop sales, a case might be made for subsidizing the difference between WTP and actuarially fair premiums, at least in some instances. A subsidy deserves further exploration and its details would have to be worked out for specific contracts and situations. More broadly, establishment of interlinked markets for input, credit, and insurance packages deserves attention as a means to promote access to credit and inputs for poor farmers.

Policy message 4: Adoption and use of modern inputs remains low. Lack of credit access, coping capacity, connectivity, and profitability limits agricultural productivity.

Adoption of modern inputs remains strikingly low in most of the case study countries, especially Ethiopia and Madagascar. In Ethiopia, the percentage of cultivated area using chemical fertilizer and improved seeds is estimated at less than 50 and less than 5 percent,[9] respectively (World Bank 2005b). In Madagascar, the use of chemical fertilizer is limited to only 6 percent of the cultivated plots and the use of improved rice seeds to 9 percent of the rice fields. Modern input use was more widespread in Kilimanjaro; nearly 80 percent of the farmers had used fertilizer in the previous year, compared with 50 percent in Ruvuma. The total value of inputs used in Kilimanjaro was estimated at about $30 per acre, compared with slightly less than $10 per acre in Ruvuma.[10]

Consider this low use of inputs in light of the case studies' production function analyses, which emphasize that increased input use is an important factor in raising agricultural output. The elasticity of gross revenue from crop production with respect to the value of modern inputs was estimated at 0.46 in Kilimanjaro and 0.14 in Ruvuma. Studies synthesized in Minten (2006) and an analysis using panel data estimation (Randrianarisoa and Minten 2006) indicate that rice yields would increase by about 6 kilograms for every additional kilogram of fertilizer.

In Ethiopia, the difference in cereal yields of fertilized and nonfertilized fields is estimated at about 200 kilograms; the largest difference is reported for maize production (table 4.3). These estimates are based on a bivariate comparison. When controlling for other production factors, the elasticity of the value of cereal output in Ethiopia with respect to fertilizer use is estimated at 0.12, similar to the results obtained for Ruvuma. Consistent with the conjectured concavity of the production function (World Bank 2006a), the elasticity decreases for farmers who have already increased their fertilizer use.

These case study findings echo what has been found elsewhere in Sub-Saharan Africa. Increased use of fertilizer, improved seeds, and integrated pest management will help raise agricultural output (World Bank 2006b). Fertilizer use is particularly important, given the wider problem of soil degradation in Sub-Saharan Africa (InterAcademy Council 2004). But an increase in the use of fertilizer or in improved seeds alone is unlikely to generate the expected and necessary boost in agricultural output. Important synergies arise from combining these inputs, as has been highlighted in an analysis of factors influencing food productivity in Ethiopia (World Bank 2005b).

A host of demand-side factors, including inability to cope with shocks and limited knowledge, profitability, and access to credit, have been invoked to explain the low adoption of modern inputs in Sub-Saharan Africa. Supply-side factors limiting the application of modern inputs include poor access to markets to sell surplus and a fertilizer supply chain that functions poorly due to high transaction costs related to poor infrastructure or to the organization of the distribution system. Poor fertilizer supply chains reduce the timely availability of affordable fertilizer in the market. The relative importance of these constraints in adopting inputs differs across the case studies.

For Tanzania, Sarris, Savastano, and Christiaensen (2006) emphasize limited access to credit as a key demand-side constraint and limited connectivity and timely and easy access to inputs as a key supply-side constraint. In Kilimanjaro and Ruvuma, only about one in six farmers have easy access to credit for seasonal inputs. But easy and timely access to credit was associated with an increase in the value of inputs of 20 percent in Kilimanjaro and 14 percent in Ruvuma. Other quantitatively meaningful factors in explaining the value of inputs use are the education level of the most educated women in the household (and in Kilimanjaro, the most educated man as well), visits by an extension agent, and possession of collateral (either animals or off-farm income).

Strikingly, the value of inputs among households in villages with a sales point for agricultural inputs was found to be 41 percent larger in Kilimanjaro and 83 percent larger in Ruvuma. Although this effect must partly capture placement effects, the coefficient is sufficiently large to suggest that easy access to inputs is crucial for fostering input use. Similarly, the value of inputs used among households in villages with a bus service, a good measure of the village's integration into the rest of the economy, is 20 to 53 percent higher in Kilimanjaro and Ruvuma. Note that the marginal effects of both easy access to inputs and connectedness are especially high in Ruvuma, where villages are much less well connected. Moreover, Ruvuma is more remote than Kilimanjaro.

Access to credit was in turn strongly associated with membership in a savings and credit organization (SACCO). SACCO members were 10 percentage points more likely to have easy access to credit for inputs. Access to irrigation facilities further improved a household's chances of having access to seasonal credit in Kilimanjaro, as did input sales points in Ruvuma. Households in Ruvuma villages with an input sales point were 37 percent more likely to have easy access to seasonal credit.

Finally, access to credit is highly associated with cultivation of tobacco (in Ruvuma), though not with coffee production. Tobacco farmers work under contract to tobacco firms, from which they receive knowledge and inputs. Markets that linked coffee growers and cooperatives or coffee curing factories disappeared after the coffee market was opened to private traders in 1995. The role of contract farming and interlinked credit-insurance-input markets in improving access to credit should be further investigated, especially in areas where cash crops are grown. How such arrangements could be designed for food crop growers, for whom enforcement of contracts is difficult, remains a challenge. The observation that access to credit has improved in Ruvuma villages with input sales points deserves further exploration in this context.

In Ethiopia, Dercon and Christiaensen (2005a) also find that having access to credit is important, though their analysis focused on households' ability to cope with shocks (either ex ante through water management or ex post through weather insurance, for example). Their study empirically confirms the existence of a risk-induced poverty trap. Limited profitability of fertilizer use also emerges as an issue in Ethiopia (World Bank 2006a, table 9), suggesting that that the crop/fertilizer price ratio is too low. Fertilizer use could be made much more profitable through more effective application and better farming practices (World Bank 2006a). Controlling for household fixed effects and transitory rainfall effects, Dercon and others (2006) estimate that receipt of at least one visit from an agricultural extension agent increases crop income and consumption growth by 15 and 7 percent, respectively, and reduces the incidence of poverty by nearly 10 percent. However, this effect might reflect the transfer of technology and farming knowledge (a demand-side constraint) as well as better access to fertilizer, which in Ethiopia is distributed through extension agents.

In Madagascar, Minten (2006) emphasizes the importance of remoteness, especially with regard to chemical fertilizer use, and to a lesser extent the advantages of irrigation (that is, risk control) in promoting input and technology adoption. Literacy is also positively associated with technology adoption. Distance to extension agents is especially critical in adoption of the Système de Riz Intensive. Minten did not find that this cultivation system affects adoption of other modern inputs or production techniques. Randrianarisoa and Minten (2006) emphasize the importance of risk and high fertilizer price (in part linked to poor infrastructure) in explaining the extremely low use of chemical fertilizer in Madagascar. They estimate that the ratio of the average marginal value of output to the cost of fertilizer between 2000 and 2004 was slightly below one in only one year, while it was close to or even over two in the other years. Yanggen and others (1998) find that the ratio should be at least two and preferably three to compensate for uncertainty and thereby facilitate adoption of fertilizer use. The ratio in Madagascar can be partially linked to high fertilizer prices. Randrianarisoa and Minten conclude that almost all farmers, independent of wealth, would use fertilizers if prices were similar to those in East Asian countries that grow rice.

In summary, limited access to credit and limited capacity to cope with shocks emerge as important demand-side constraints (World Bank 2006b). Institutional arrangements such as contract farming to link insurance, credit, and input markets, as well as institutional devices to promote saving (for example, SACCOs), deserve investigation, because they could facilitate access to credit. On the supply side, reduction of fertilizer sourcing costs (by lowering trade barriers to enable fertilizer importers to capture economies of scale), reduction of fertilizer distribution costs (through better rural and rail road infrastructure), better business finance and risk management instruments, and better supply chain coordination mechanisms (through introduction of standards and grades and market information systems) should be pursued to improve the crop/input price ratios.

Policy message 5: Payoffs to improved rural road infrastructure are substantial and so are the costs, underscoring the need for careful and comprehensive cost-benefit analyses.

Remoteness and access to roads emerge in all country case studies as important impediments to increasing agricultural productivity. Roads facilitate access to inputs and open markets for selling outputs by reducing transaction costs. They also improve access to off-farm employment. The empirical evidence from Ethiopia regarding the responsiveness of agricultural output, crop income, and consumption/poverty is illuminating in this regard. Using their 1994–97 panel of 1,500 rural households from 15 villages in Ethiopia, Abrar, Morrissey, and Rayner (2004) find a strong effect of market accessibility (as captured by the population size of the nearest town divided by the road distance to that town), though mainly for the two most widely traded crops (maize and teff) and especially in more densely populated and better connected areas. Extending the same panel of households with three more observations per household to 2004, Dercon and others (2006), find a meaningful, positive effect of access to an all-weather road on growth in crop income. Such access increased consumption growth by 16 percent and reduced the incidence of poverty by 6.7 percent.

Similarly, using quintiles of a remoteness index derived through factor analysis of proxies of access to markets and services, Randrianarisoa and Minten (2006) find that residence in better-connected areas facilitates technology adoption in Madagascar. But as illustrated in Christiaensen (2004), at $300,000 per kilometer of paved trunk road, the costs of constructing roads, rehabilitating them, or both in rugged areas such as Ethiopia (and Madagascar) can be substantial, underscoring the need for careful cost-benefit analysis, especially when it concerns sparsely populated areas.

Policy message 6: The constraints faced by poorer farmers are not necessarily those faced by richer farmers.

One of the advantages of micro-level data is that they allow different types of households to be distinguished one from the other. The Madagascar study by Randrianarisoa and Minten (2006) is particularly interesting from this perspective. It found important differences in the yield constraints facing poor and nonpoor rice farmers:

- Halving the average age at which rice is transplanted (a proxy for use of modern technology) increases yields by 13 percent on average, but

by 18 percent for the poorest two quintiles (Randrianarisoa and Minten 2006, table 5).
- Doubling fertilizer use would increase yields by 8 percent for the poorest quintile but by only 5 percent for the richest quintile.
- Doubling the number of cattle owned would raise rice yields of the richest farmers by 13 percent and those of poorest farmers by only 5 percent.
- Doubling ownership of agricultural equipment would raise yields by 5 percent for the richest quintiles, but by only 1 percent for the poorest quintiles.
- Rainfall shocks are the severest constraint for the poorest farmers. They reduce yields of the poorest quintile by 35 percent and yields of the richest quintile by 10 percent.

Encouraging the adoption of new technology, including the use of fertilizer, would clearly benefit poor farmers (as would raising their educational achievements). Insurance against weather risks is another intervention that would appear to be pro-poor in Madagascar. Nonetheless, only 29 percent of farmers in the poorest income quintile in Madagascar would be willing to pay 100 Ariary per year to operate and maintain an irrigation system (Randrianarisoa and Minten 2006, 29); 42 percent of farmers in the richest quintile would pay that amount. But poor farmers would benefit less from interventions that target ownership of cattle and agricultural equipment.

In conclusion, the cross-country macroanalysis presented above shows that the historically observed lower growth in agriculture compared with growth in other sectors should not be interpreted as proof that agriculture has low productivity growth potential. On the contrary, total factor productivity in agriculture has historically increased at least as fast as that in other sectors. Although this phenomenon does not exclude a passive contribution from agriculture to other sectors through the release of surplus labor, it does suggest that agricultural productivity and growth can be substantially increased and that they must be increased to facilitate migration of labor from agriculture (Gollin, Parente, and Rogerson 2002) and foster nonagricultural growth (as discussed in chapter 5).

The finding from the micro data that Sub-Saharan farmers produce on average 30 percent less (or about 40 percent in the case study countries) than best practice among their peers in the region confirms that substantial scope for growth in agriculture remains, especially given that the current production frontier usually lies well below the frontier achieved in experimental stations. A package of interventions carefully tailored to local circumstances could help close the efficiency gap and expand the production frontier. Entry points for public intervention include continued emphasis on decentralized agricultural research and extension responsive to the needs of farmers, provisions adequate to help farmers cope with risks and shocks (such as better water management and rainfall-based insurance schemes), institutional arrangements to foster smallholder farmers' access to credit (such as contract farming and mechanisms to promote saving), public investment in rural infrastructure, and policy and institutional reforms to improve access to and reduce the costs of modern inputs. The relative importance to be accorded to each of these areas (and others) will depend on the local context.

Notes

1. This section is largely taken from Christiaensen, Demery, and Kuhl 2006.
2. See North (1959), Johnston and Mellor (1961), Hayami and Ruttan (1985), and Timmer (1997).
3. The powerful dual economy models of development, for example, critically assume a stagnant rural sector from which labor and resources flow to the dynamic, modern industrial sector.
4. Productivity estimates for the economy as a whole or for individual sectors abound.
5. GDP is denoted by G_k, and population in sector k is denoted by L_k. Through total differentiation of $G_k = L_k (G_k/L_k)$, it can readily be shown that

$$\frac{dG_k}{G_k} = \frac{d(G_k/L_k)}{(G_k/L_k)} + \frac{dL_k}{L_k}.$$

6. Sherlund, Barrett, and Akinwumi (2002) have highlighted the importance of controlling for soil quality in the estimation of technical efficiency. Failure to do so may lead to an estimated inefficiency higher than actual inefficiency.
7. SRI is a labor-intensive methodology to cultivate rice. It relies on adjustments to traditional agronomic practices and little use of external inputs. SRI has been shown to increase yields from an average of 2 tons per hectare to 6 or more tons per hectare. But adoption rates have remained low, and abandonment rates have been high due to lack of continuous agricultural extension, deficient water management, seasonal liquidity constraints, and increased yield risks (Minten 2006).
8. The technical inefficiency estimates found in the case studies appear somewhat above the average of 30 percent technical inefficiency reported in the

literature (Ali and Byerlee 1991). Binam and others (2004) report an average technical inefficiency among smallholder farmers in the slash-and-burn zone of Cameroon of 25 percent. Kaliba (2004) finds that yields of smallholder dairy farmers in central Tanzania are on average 30 percent below the production frontier. Ajibefun, Battese, and Daramola (2002) estimated that food crop farmers in Oyo State were on average 18 percent inefficient in their use of production inputs.

9. These estimates are based on the official nationally representative agricultural sample surveys and may be slight underestimates, given definitions of improved seeds.

10. The lower current use of fertilizer and other modern inputs in Kilimanjaro and Ruvuma is likely to be linked to the extremely low coffee prices during the years of the survey.

CHAPTER 5

Agriculture and Growth in the Rest of the Economy

Agricultural development not only has a direct effect on income and economic growth, but also plays an important role in fostering development in nonagricultural sectors of the economy (Johnston and Mellor 1961; Schultz 1964). Three broad types of development mechanisms have been identified. The first is *intersectoral links*, forward to agroprocessing and backward to input supply sectors (see Perry and others 2005). The second mechanism is *final demand effects* arising from a large and growing agricultural sector with a propensity to spend on locally produced nontraded goods and services (especially true of smallholder agriculture), thereby generating significant demand for nonagricultural goods (see Haggblade, Hammer, and Hazel 1991) and creating off-farm employment. The third mechanism is *wage-goods effects*—by reducing the price of food, agricultural productivity growth lowers the real product wage in nonagricultural sectors, thereby raising profitability and investment.[1]

Much of the literature has argued that the stronger links are from agriculture to other sectors, rather than the other way around (Mellor 1976; King and Byerlee 1978; Thirtle and others 2001), because inputs into nonagricultural sectors are more import intensive, and urban consumption patterns favor imported goods (the demand for food being income inelastic). As Timmer (2005) points out, establishing the empirical validity of these links has been a cottage industry since the early 1970s.

The models adopted in the literature typically embrace both production and consumption linkages,[2] but the latter have been found to be more important. Delgado, Hopkins, and Kelly (1998) concluded that for both Africa and Asia, consumption-based agricultural growth links are four to five times more important to growth than production-based links. For the linkage effects to be significant, four conditions must apply (Delgado, Hopkins, and Kelly 1998). First, agriculture must be a sufficiently large sector in employment terms for income-generating effects to be significant. Second, the income gains from agricultural growth must be reasonably widespread, so that effective demand for locally produced goods and services increases. Third, the consumption patterns of people in agriculture must favor locally produced nontradable goods and services. And finally, the nonagricultural (nontraded) sectors must have underutilized resources and appropriate institutional arrangements to be able to respond to increased levels of demand from agricultural growth.

Using micro data on consumption patterns in five African countries, Delgado, Hopkins, and Kelly concluded that the farm sector in Africa is better able to propagate income growth than previously thought: "adding $1.00 of new farm income potentially increases total income in the local economy by an additional $1.88 in Burkina Faso, $1.48 in Zambia, $1.24 to $1.48 in two locations in Senegal, and $0.96 in Niger" (Delgado, Hopkins, and Kelly 1998, xii). After accounting for the potentially inelastic supply response of the nontradable nonagricultural sector, Delgado, Hopkins, and Kelly estimate the agricultural multiplier effects to be around 1.1 for African countries but 1.6 for Asian countries. This difference is ascribed to Asian economies' abundance of labor, which increases the supply response of the Asian nontradable sector. Similarly, by applying computable general equilibrium (CGE) models to archetype economies for Africa, Asia, and Latin America, de Janvry and Sadoulet (2002) find the employment and overall linkage effects from increased land productivity in agriculture to be more important in Asia and Latin America, where labor and food markets are relatively well developed, than in Africa.

Dorosh and Haggblade (2003) confirm the existence of sizeable growth links from investments in agriculture by applying fixed-price semi-input-output (SIO) models and fully price-endogenous CGE models to eight Sub-Saharan African countries. As before, the indirect effects of investment-induced growth prove to be about as large as the direct effects. Moreover, fixed-price (SIO) multipliers from investments in export and food crops typically exceed manufacturing multipliers, consistent with the literature, though not when prices of the nontradables are endogenized (as in the computable general equilibrium).

The methodologies—fixed-price (semi) input-output models and price-endogenous CGE models—underpinning this micro evidence are structural in nature and data intensive. Thus the methodologies provide many insights into the nature and extent of linkage effects, but their results (partly) depend on the validity of their structural assumptions. Bravo-Ortega and Lederman (2005) looked for evidence of a causal relationship between agricultural and nonagricultural output (in the sense proposed by Granger 1969[3]) by applying dynamic panel data estimation techniques to cross-country data. In this way they did not, for example, have to assume supply flexibility or fixed prices in the nontradable sector, in effect observing a "reduced form" of the full general equilibrium outcome. The downside of this approach is that it provides little insight into the mechanisms of linkage effects.

Bravo-Ortega and Lederman find evidence of a positive causal link (in the Granger sense) between nonagricultural growth and (lagged) agricultural growth for poor countries; the linkage effect is smallest for Latin America. They also discern a lagged effect of nonagricultural output on agriculture, though this effect is negative for low-income countries and only slightly positive (and not statistically significant) for Latin American countries. Although both the micro data and this cross-country evidence suggest that the linkage effects of agricultural growth appear to hold for Africa, the recent opening up of African economies might undermine the linkage effects from any expansion in rural demand, which might increasingly be met by imports.

Christiaensen, Demery, and Kuhl (2006) update the data set examined by Bravo-Ortega and Lederman to cover the period 1960–2004 and revisit their results taking an African focus and using a somewhat modified specification (including yearly deviations from long-run average rainfall in the agricultural regressions to ensure identification of the model in the long run). They find a small positive effect of past growth in agriculture on growth in the non-agricultural sector, though only in the low income countries (excluding Sub-Saharan Africa). They do not find evidence of a reverse effect (from growth in nonagriculture to agriculture). These results are consistent with those reported by Tiffen and Irz (2006) who also perform a Granger causality test, this time in a

Vector Autoregressive (VAR) model using bootstrapping methods to improve the small sample performance of the test statistic. They conclude that agricultural value added per worker (Granger) causes GDP growth per capita in most developing countries, while the direction of causality in developed countries is unclear.

This volume's four case studies rely mostly on cross-section micro data, and most do not seek to assess the indirect effects of agricultural growth on the nonagricultural economy beyond effects on prices and wages. Only the Madagascar case study used CGE models to assess the economywide effects of investments in agriculture and nonagricultural sectors (Randriamiarana, Razafimanantena, and Minten 2006). It concludes that in a general equilibrium setting, improvements in total factor productivity in the staple sector (rice) have major effects on urban poverty—a result of the fall in staple prices in the wake of productivity gains. The rural poor also benefit, but their gain is muted, because the income gains from the increase in productivity are reduced to some extent for net sellers by the fall in output price. A similar productivity gain in the cash crop sector would have less impact on poverty, because such a change would affect fewer people. With respect to the nonagricultural investments examined, a reduction in marketing margins would help the rural poor, whereas a doubling of the tourist sector or investments in export processing zones would mainly help the urban population.

To conclude, the micro evidence from structural models discussed in the literature, the cross-country regressions, and the CGE analysis in Madagascar indicate that the indirect effects of fostering growth in agriculture are on average substantial, even though they tend to be weaker in Sub-Saharan Africa than in Asia and Latin America. Although some recent evidence calls into question whether agricultural multipliers largely exceed nonagricultural multipliers (Dorosh and Haggblade 2003), virtually all studies concur that the feedback effects from agriculture to nonagricultural sectors are on average at least as large as the reverse effects. Finally, looking beyond the averages, the exact magnitude of the sectoral multipliers is likely to depend on a given country's institutions and economic structure.

Notes

1. Lower food prices also raise real consumption wages and thus directly benefit poor (urban and rural) wage earners, as discussed in the analysis of the determinants of participation effects in chapter 3.

2. Production links refer to intersectoral purchases and sales of intermediate goods. Consumption links occur when the incomes generated by growth in one sector lead to increases in final demand for the goods of other sectors and as a result increase employment in those sectors.
3. The concept of Granger causality holds that a variable Y_{ai} Granger causes Y_{ni} if Y_{ni} can be better predicted using lagged values of Y_{ai} than without them.
4. To be consistent with the classification applied in the analysis of the participation effect in chapter 3, the same cutoff value of $1,160 was used.

CHAPTER 6

A Sectoral Decomposition of Poverty Change

Equation (5) in chapter 2 and equation (2) in chapter 3 decomposed poverty reduction during a certain period into sectoral participation and growth effects. Chapter 3 provided estimates of the participation effects of the different sectors and concluded that 1 percent per capita agricultural growth yields on average 1.6 times more (headcount) poverty reduction than 1 percent per capita growth in industry and 3 times more than 1 percent per capita growth in the services sector. Chapters 4 and 5 showed that agriculture is likely to continue to grow at a slower pace than nonagricultural sectors due to Engel's Law, but that the indirect effects of agricultural growth on other sectors are substantial, especially in closed low-income economies, and at least as large as the reverse feedback effects. Whether investments in agriculture in a particular country would, as a result, generate faster or slower overall economic growth than investments in nonagricultural sectors is a priori unclear. The answer would depend on the country's economic structure and institutional arrangements.

How do the potentially counteracting forces of relatively slower overall growth from investments in agriculture and agriculture's relatively larger participation effect play out for poverty reduction? To answer this question, the relative contributions of the agriculture, industry, and services sectors to the (predicted) change in ($1/day) poverty incidence are calculated using the estimates of equation (2) reported in chapter 3. The estimated coefficients from pooled data set reported in table 3.2, column D, are applied to observed sectoral GDP growth rates in the PovCal sample for the years 1980 to 2000. Approximately two-thirds of the poverty periods reflected in the sample occur in the 1990s.

Examination of the relative contribution of agriculture during this period is useful, because that contribution coincides with increasing liberalization and globalization of the world economy. These evolutions might affect the feedback effects of agriculture to nonagricultural sectors, as well as the participation effects, if globalization increased the correlation between domestic and international food prices. The effect of service growth and of the constant (a measure of the effect of inequality change) is retained in the decomposition, even though the estimated coefficient of each is not statistically significant. As noted in chapter 3's discussion of robustness tests, the imprecision in the estimated coefficient on service growth is likely to be related to the lower signal to noise ratio as all observations are included. When (counterintuitive) spells with substantial positive (negative) growth rates and substantial poverty increases (decreases) are omitted, growth in services becomes statistically significant, without meaningful changes in the size of the coefficients. Nonetheless, the study opted for minimal inference in the data set.

Table 6.1 presents the results of the sectoral decomposition of headcount poverty for the lower-income countries, the higher-income countries, and the pooled data set. The median value of $z/\bar{e}$ (=0.246) is used to classify spells in the lower-income and higher-income categories. Given the limited sample size, spells in which the share of a sector to poverty reduction falls beyond the [−3, 3] range are excluded to obtain more stable results; a sample of 232 observations is left. Using the [−5, 5] or [−1, 1] range yields qualitatively similar results, although the latter cutoff substantially reduces the sample. Using the regression coefficients for the pooled sample, the poverty headcount is predicted to fall by 4 percent on average, compared with an observed reduction of 5 percent.[1] Growth in agriculture contributed as much to this reduction—about one-third—as growth in industry, despite the fact that the share of agriculture in the total economy was on average only 19 percent, compared with 32 percent for industry.

The contribution of agriculture to poverty reduction is a bit lower in the higher-income countries, where the agricultural sector is also smaller. About one-quarter of the change in poverty is attributed to the

Table 6.1. Sectoral Decomposition of Changes in Headcount Poverty

Average contributed share to poverty reduction of	Number of observations	Observed % change in poverty headcount	Predicted % change in poverty headcount	Agriculture	Industry	Services	Inequality change
Lower-income countries							
Sub-Saharan Africa	31	−0.07	0.04	**0.57**	**0.15**	−0.02	0.29
South Asia	20	−0.13	−0.12	**0.23**	**0.20**	−0.18	0.76
East Asia and Pacific	36	−0.24	−0.18	**0.22**	**0.38**	0.09	0.32
East Europe and Central Asia	10	0.11	0.40	**0.69**	**0.29**	0.01	0.00
Latin America and the Caribbean	18	−0.02	−0.25	**0.43**	**0.22**	0.02	0.32
Middle East and North Africa	2	−0.27	−0.13	**0.19**	**0.23**	0.20	0.38
Total	117	−0.11	−0.07	**0.39**	**0.25**	0.00	0.36
Higher-income countries							
Sub-Saharan Africa	2	−0.13	−1.01	**0.21**	**0.09**	0.09	0.61
East Asia and Pacific	9	−0.22	−0.30	0.02	**0.66**	0.25	0.07
East Europe and Central Asia	37	0.33	0.24	**0.14**	**0.43**	0.24	0.19
Latin America and the Caribbean	58	−0.11	−0.02	**0.28**	**0.43**	0.10	0.19
Middle East and North Africa	9	−0.11	−0.34	**0.64**	**0.14**	0.16	0.06
Total	115	0.02	0.00	**0.24**	**0.42**	0.16	0.18
All countries (pooled sample)							
Sub-Saharan Africa	33	−8	−3	**0.55**	**0.15**	−0.01	0.31
South Asia	20	−13	−12	**0.23**	**0.20**	−0.18	0.76
East Asia and Pacific	45	−23	−21	**0.18**	**0.43**	0.12	0.27
East Europe and Central Asia	47	28	27	**0.26**	**0.40**	0.19	0.15
Latin America and the Caribbean	76	−9	−7	**0.32**	**0.38**	0.08	0.22
Middle East and North Africa	11	−14	−30	**0.56**	**0.15**	0.17	0.12
Total	**232**	**−5**	**−4**	***0.31***	***0.34***	***0.08***	***0.27***

Source: Authors.
Note: Boldfaced shares are based on statistically significant coefficients.

performance of the agricultural sector, even though it comprises only 13 percent of GDP on average. In the lower-income countries, however, about 40 percent of the change in poverty is attributed to the performance of agriculture.

In Sub-Saharan Africa, agriculture is responsible for 60 percent of the predicted evolution of poverty. The poverty increase in many Sub-Saharan African countries in the sample implies that the lack of performance in the agricultural sector has largely contributed to the region's overall lack of welfare improvement (and to a welfare decline in many countries). Average annual per capita agricultural GDP growth in the African spells of the low-income sample was estimated at –0.2 percent. This result suggests that insufficient attention to the needs of the agricultural sector, rather than lack of participation by the poor in agriculture, has led to adverse poverty outcomes in the region. The dismal performance of the agricultural sector in the lower-income countries of Eastern Europe and Central Asia, where average annual per capita agricultural GDP growth is estimated at –0.8 percent, is also an important factor in explaining the poverty increase in these countries.

The above-noted estimates are likely to be lower bounds on agriculture's contribution to poverty reduction. Given that the decompositions use contemporaneous sectoral growth rates, they do not account for the contributions of growth in agriculture to growth in other sectors, which are usually larger than the reverse feedback effect. In summary, despite increasing liberalization and globalization, agricultural growth has contributed substantially more than its share in the economy to poverty reduction in recent history. It contributes most to poverty reduction in the low-income countries, and its performance, or lack thereof, has proven to be especially important in understanding the evolution of poverty in Sub-Saharan Africa.

Note

1. The average predicted poverty incidence for the total sample equals the average observed poverty incidence as expected from theory.

CHAPTER 7

Concluding Observations

The contribution of a sector to poverty reduction depends on its participation effect and growth effect. The former is determined by the elasticity of total poverty to sectoral growth and the sector's share in the overall economy. The latter has both direct and indirect components. The participation effect of agricultural growth on poverty is on average about 1.6 to 3 times larger than that of growth in other sectors. The larger participation effect of growth in agriculture does not primarily follow from the large share of agriculture in national economies, but rather from its larger GDP elasticity of overall poverty. That agricultural growth is more effective in reducing poverty holds true in lower- and higher-income countries, though the comparative advantage of agriculture in reducing poverty declines as countries become richer. At −1.8, the participation effect of agriculture is the largest in Sub-Saharan Africa and over 3 times larger than the effect on poverty incidence of an additional percentage point of growth in other sectors.

A review of the evidence on the direct growth contributions of agriculture and nonagricultural sectors reveals that the agricultural sector has on average grown 1 to 3 percentage points more slowly than nonagricultural sectors. However, this lower rate of growth appears largely due to a migration of labor from agriculture into other sectors and is not necessarily a consequence of slower productivity growth in agriculture. The migration of labor out of agriculture could indeed be in response to more productive opportunities in nonagricultural sectors (industrial pull). However, the more rapid increase in agricultural total factor productivity that historically has been observed in both developed and developing countries lends support to the view of agriculture as a dynamic sector with substantial growth potential. Productivity increases in agriculture induce labor to move out of agriculture (agricultural push). While both industrial pull and agricultural push have likely been at work in recent decades, the key conclusion of the evidence reviewed here is that agriculture is not a backward sector with inherently inferior productivity growth potential and that much scope for improving agricultural growth remains.

Nonetheless, agricultural growth is likely to continue to lag behind growth in nonagricultural sectors due to Engel's Law. Agriculture's direct growth effect is thus likely to be smaller than that of other sectors. But both the micro and the cross-country evidence indicate that the indirect or growth linkage effects from agriculture to nonagricultural sectors are quantitatively large and at least as large as the reverse feedback effect. Although the evidence suggests that these indirect growth effects are likely to be smaller in Sub-Saharan Africa than in the rest of the developing world, they appear nonetheless quantitatively important, and they may well compensate for the smaller direct growth effect of agriculture.

In summary, the empirical evidence suggests that the larger participation and indirect growth effects of agriculture are likely to offset the sector's smaller direct growth effect, especially in the lower-income countries, including those in Sub-Saharan Africa. Decomposition of poverty incidence periods from the 1980s and 1990s into their sectoral contributions supports the view that agriculture plays a critical role in the evolution of poverty, especially in low-income countries. Agriculture explained on average about 40 percent of the predicted poverty change, and its dismal performance explained more than half of the observed stagnation or increase in poverty in Sub-Saharan Africa. In higher-income countries, agriculture contributed relatively more to poverty reduction when compared with its share in the economy than other sectors. These results were obtained in a period of increasing liberalization and globalization, both of which potentially undermined agriculture's indirect growth and participation effects. Nonetheless, the effect of liberalization and globalization on the advantage agriculture holds over other sectors in reducing poverty deserves continued empirical attention.

The evidence presented in this volume supports the view that enhancing agricultural productivity is a critical starting point in designing effective poverty reduction strategies, especially in the low-income countries. It also highlights the need to look beyond the averages. Partial equilibrium analysis shows that participation of the poor in agricultural growth in the wake of an increase in total factor productivity is likely to vary depending on the net food marketing position of the poor (net seller/buyer) and on the product price and labor market changes induced by changes in agricultural productivity. The latter effects depend in turn on the nature of the internal and external trade regimes and on the price elasticity of demand. Net food sellers may lose from increases in agricultural productivity, depending on how price inelastic the demand for the produced good is and on whether the decline in its wage bill is sufficient to compensate for the potential decline in revenues from its output in the wake of declining prices. Net buyers usually stand to gain, unless the potential reduction in wage income would exceed the decrease in output prices.

Contrary to what is commonly assumed, many of the rural poor are actually net food buyers, even in very poor countries in Sub-Saharan Africa. Many of the rural poor (in addition to the urban poor) would thus stand to gain from an increase in agricultural productivity. Net food sellers, however, are usually farmers with large-scale operations, and a small minority of farmers produce most of the marketed surplus. This reality explains why the maize market interventions by the National Cereal Producers Board in Kenya were regressive. These interventions reduced the volatility of maize prices but also increased those prices by 20 percent. Because 83 percent of the rural households in the lowest income quintile are net food buyers, these market interventions actually increased poverty and hit the very poor hardest. Abandonment of the interventions would increase these households' income by 18 percent.

If raising agricultural productivity holds much promise to reduce poverty, how should such productivity be increased, especially in Sub-Saharan Africa, where it has been lagging? The four country case studies reviewed in this volume highlight several policy messages pertinent to this question. First, the studies emphasize that (especially, though not only) staple crop farmers in many African countries continue to produce well within the production frontier—that is, well below their best-practice counterparts because of technical and allocative inefficiency. Moreover, agricultural research suggests that a substantial outward shift of the production frontier should be well within reach. In other words, given current agronomic best practice and available knowledge, there appears substantial scope for increasing agricultural productivity.

Second, the determinants of farm efficiency and productivity are complex, and their relative importance varies substantially by location. This reality underscores the need for location-specific (and comparative) analysis as well as a decentralization of research and extension activities. Too many poverty reduction strategies are developed without a proper understanding of the basic facts of the given agricultural sector, the rural economy and its links with the rest of the economy, and the implications of those facts for increasing the effect of agricultural development on poverty.

Third, most agriculture in Sub-Saharan Africa will continue to be rainfed, and both the immediate and long-term impact of weather-related shocks is substantial, as illustrated by the case study evidence from Ethiopia, Madagascar, and Tanzania. In addition, the Ethiopian case study provides one of few robust empirical studies that farmers' inability to cope with shocks ex post induces them to forgo adoption of mean-enhancing technologies such as fertilizer. According to that study, an increase in the level of consumption when the harvest fails by one standard deviation of the cross-sectional distribution of consumption will increase fertilizer use by 44 percent. The study should encourage expansion of irrigation and exploration of the merits of weather-based insurance in different settings. The empirical analyses confirm the substantial benefits of and demand for such interventions, but the costs are often considerable. Therefore, cost-benefit analyses should inform the choice of appropriate risk-mitigating instruments.

Fourth, in addition to poor agronomic practices, limited adoption of modern inputs such as fertilizer, improved seeds, and integrated pest management leads to inefficient crop production. The Ethiopian case study illustrates the importance for fertilizer adoption of farmers' ability to cope with shocks. Another determinant of inefficiency is limited access to credit, an important constraint to adoption of fertilizer in Madagascar and Tanzania. Provision of weather-based insurance to credit-providing institutions should be explored, together with institutional arrangements that encourage farmers to save to overcome prevailing credit constraints. Better water management (water harvesting and irrigation) will not only help reduce risks, but also help increase the use of inputs. The Tanzania case study indicated that agricultural extension would increase adoption of inputs, and the Ethiopia case study indicated that it would increase crop income.

Concluding Observations

Finally, limited market access and limited profitability of input use are key supply-side impediments to increasing agricultural productivity. The benefits of improving road access to increase market access are most often substantial, but so are the costs. Cost-benefit analyses of rural road investments are particularly needed when the investments are considered for rugged and sparsely populated areas. In conducting such analyses, the debate about investment in agriculture versus investment in other sectors is often misleading, because many public investments (especially in rural roads but also in education and health services) are as necessary for developing the nonfarming sector as they are for developing the farming sector.

Appendix 1
Objectives of and Data Sources for Case Studies

Country	Objective of study	Data set used
Ethiopia		
Dercon and Christiaensen (2005a)	To explain low use of fertilizer input arising from rainfall uncertainty, thus leading to poverty traps	Ethiopia Rural Household Survey (ERHS), a longitudinal household data set covering 1,477 households in 15 peasant associations across the 4 major regions in Ethiopia; surveyed five times between 1994 and 1999
Dercon and Christiaensen (2005b)	To show that fertilizer use favorably affects household consumption and poverty reduction	ERHS; surveyed five times between 1994 and 1999
Kenya		
Jayne, Meyers, and Nyoro (2005)	To assess policy interventions' impact on maize prices	Monthly data from January 1989 to November 2004 on the wholesale maize market prices for Kitale and Nairobi; obtained from Ministry of Agriculture's Market Information Bureau

(continued)

Objectives of and Data Sources for Case Studies (continued)

Country	Objective of study	Data set used
Mude (2005)	To estimate the welfare effects of the policy-induced maize price hike	Two cross-sections of survey data, collected in 1997 and 2000, each consisting of about 1,500 rural farm households surveyed by the Tegemeo Institute; 1997 Welfare Monitoring Survey to assess effect on urban households and urban poverty
Mistiaen (2006)	To estimate the extent of maize production inefficiency and to assess its implications for poverty in Kenya	Two cross-sections of survey data, collected in 1997 and 2000, each consisting of about 1,500 rural farm households surveyed by the Tegemeo Institute; agronomic field experiment data collected from 1987 to 1993 through the Fertilizer Use Recommendation Project set up by the Kenya Agricultural Research Institute
Madagascar		
Minten, Randriamiarana, and Razafimanantena (2006)	To assess the role of agriculture in influencing poverty dynamics during the 1990s and to assess the impact of gains in agricultural productivity on poverty	Three rounds (2000–03) of a specially designed panel survey in two highland regions covering 237 households
Minten and Barrett (2006)	To estimate the impact on nutrition-based welfare of changes in agricultural productivity and other agricultural factors	A specially conducted 2001 census of 1,381 communes
Randrianarisoa and Minten (2006)	To identify the main determinants of agricultural (rice) productivity in two highland areas	A series of national household surveys conducted by Institut National de la Statistique Malgache in 1993, 1997, 1999, 2001, and 2002
Tanzania		
Sarris, Savastano, and Christiaensen (2006)	To assess determinants of agricultural productivity using production functions and the stochastic production frontier approach	Representative survey of about 900 smallholder agricultural households in 45 and 36 villages in Kilimanjaro and Ruvuma region, respectively; surveyed in November 2003 and February 2004

APPENDIX 2

Data Sources and Constructs Used in Case Studies

Empirical approaches to assessing the link between agricultural productivity and poverty take three forms: econometric estimation based on country data (for example, the studies by Ravallion and Chen 2007 and Ravallion and Datt 1996, 1999, and 2002), model-based assessments with some micro underpinnings (such as Delgado, Hopkins, and Kelly 1998 and Fan, Chan-Kang, and Mukherjee 2005), and cross-country estimation (Bravo-Ortega and Lederman 2005). Given the importance of country-specific conditions in determining this link, and the structural limitations of modeling work, the use of country micro data is the preferred approach. The four African case studies commissioned for this volume were designed to take full advantage of country micro data sets that cover both poverty and agriculture.

The Madagascar case studies have the most compelling and varied information base for assessing the role of agriculture in poverty reduction. Perhaps the best starting point is the study by Minten and Barrett (2006), which used spatially explicit data (mainly obtained through focus group discussions) from a specially conducted census of Madagascar's 1,392 communes (1,381 were enumerated) in 2001.[1] Such a community-level spatial analysis in the African setting is rare, if not unique. Minten and Barrett's estimation specification also used Madagascar's official 1993 census and other secondary data. The researchers measured poverty as the extent of malnutrition in the community—specifically, the percentage of households perceived to be food insecure in each commune and the average length of households' lean periods.

Minten, Randriamiarana, and Razafimanantena (2005) used more orthodox data in their assessment of the poverty-agriculture link in Madagascar. They used micro data from a series of national household surveys conducted by the Institut National de la Statistique Malgache during the period 1993 to 2002 to describe the main characteristics of poor households. The surveys obtained detailed expenditure information (from which consumption poverty could be readily estimated), but because coverage of agriculture was limited (apart from the 2002 survey), the researchers were obliged to use a computable general equilibrium (CGE) model.

In the third Madagascar study, Randrianarisoa and Minten (2005) conducted three rounds of a specially designed panel survey in two highland regions during the 2000–03 period. The researchers conducted each survey three to four months after the main rice harvest. They randomly selected villages from a stratified sample to ensure sufficient coverage of villages by size and access to paved roads. They complemented their quantitative data with survey respondents' perceptions. Specifically, they asked farmers in 237 households about their willingness to pay for agricultural inputs.

The study on Tanzania by Sarris, Savastano, and Christiaensen (2006) is also based on a household survey designed to be representative of smallholder farming (including coffee and noncoffee production) in Kilimanjaro and Ruvuma, two regions differing in their agroecological and socioeconomic characteristics. The survey included about 900 households in each region. In addition to the household survey, the researchers conducted a community survey involving village focus groups. The surveys obtained both detailed agricultural production and household consumption data, the latter being the basis of poverty measurement. In their analysis of poverty, Sarris, Savastano, and Christiaensen faced two issues. The first was the difficulty of measuring self-produced food consumption, which turned out to be particularly sensitive to the method of collection—depending on whether the consumption or production module of the questionnaire was used. The researchers selected the consumption-based approach, which was consistent with other data (especially that from national surveys of household budgets). The second issue was the choice of poverty line to apply to the consumption-based welfare measure. The researchers chose the 2000/01 Household Budget Survey poverty line inflated to the month and year of the survey using the Consumer Price Index.

Mistiaen (2005) creatively combined experimental farm trial data with household survey data in assessing the extent of technical inefficiency in Kenyan maize production. He used two cross-sections (each consisting of about 1,500 rural farm households) of survey data collected in 1997 and 2000 by the Tegemeo Institute in collaboration with Michigan State University.[2] These surveys were designed to obtain representative baseline data on maize production (at the plot level) and household characteristics relevant to the 1996–97 and 1999–2000 growing seasons. The surveys cover Kenya's nine principal geographic agroecological zones (AEZs). Mistiaen combined the survey data with agronomic field experiment data collected from 1987 to 1993 through the Fertilizer Use Recommendation Project (FURP) set up by Kenya Agricultural Research Institute. The FURP conducted trials at 71 sites (in 31 districts) that are representative of the different AEZs and soil types. By matching these two data sets, Mistiaen was able to compare the yields achieved in the field trials with those obtained on farms (as reported in the Tegemeo survey). The Tegemeo data were less than ideal for assessing the impact of the estimated maize production inefficiency on poverty, because they did not include good household consumption data (the survey being designed to focus on agricultural production and income). To assess the poverty impact, Mistiaen was obliged to draw on the poverty measures obtained from the Welfare Monitoring Survey (WMS).

Jayne, Myers, and Nyoro (2005) used monthly time series data on maize prices from the National Produce and Crop Board to assess the impact of policy interventions in the Kenya maize market. They applied their results to the Tegemeo farm household survey data by Mude (2005) to assess the likely poverty impact. Like Mistiaen (2005), Mude had to draw on the 1997 WMS data to obtain poverty estimates from the Tegemeo survey. Mude selected the poverty line using the income metric of the Tegemeo data to yield the incidence of overall rural poverty that was obtained in the 1997 WMS.

The case study on consumption risk, technology adoption, and poverty traps in Ethiopia by Dercon and Christiaensen (2005a, 2005b) required panel data. The researchers used the Ethiopia Rural Household Survey (ERHS), which is a rare (in Africa at least) longitudinal household data set. The ERHS covered 1,477 households in 15 peasant associations across the four major regions in Ethiopia. These households were surveyed five times between 1994 and 1999. The sample was selected to be broadly representative of the main farming systems in the country. The researchers obtained observations for about 88 percent of the sample in all five years—an attrition rate of only about 3 percent per year.

Notes

1. The commune census was conducted under the Ilo program of Cornell University and in collaboration with INSTAT (the national statistical institute) and FOFIFA (the national center for agricultural research).
2. These surveys were implemented under the Kenya Agricultural Marketing and Policy Analysis Project, a collaboration of Egerton University/Tegemeo Institute, Michigan State University, and the Kenya Agricultural Research Institute.

Appendix 3
Decomposition of Changes in the Poverty Headcount before and after 1995

	Full sample		Spell starting before 1995		Spell starting in 1995 or thereafter	
% change in $1/day poverty gap squared poverty	Coeff.	p-value	Coeff.	p-value	Coeff.	p-value
Agricultural GDP/cap growth	−10.03	0.000	−11.34	0.000	−10.21	0.023
Agricultural GDP/cap growth × (poverty line/average income)	9.76	0.000	16.38	0.000	10.56	0.024
Industrial GDP/cap growth	−4.02	0.000	−4.17	0.000	−3.85	0.057
Industrial GDP/cap growth × (poverty line/average income)	4.38	0.005	7.94	0.000	2.41	0.400
Service GDP/cap growth	−2.02	0.141	−2.55	0.198	−1.70	0.412
Service GDP/cap growth × (poverty line/average income)	3.45	0.084	2.92	0.93	3.96	0.241
Constant	−0.06	0.182	−0.98	0.16	−0.76	0.252
# of observations/R^2	265	0.27	103	0.52	162	0.16

(continued)

Decomposition of Changes in the Poverty Headcount before and after 1995 (continued)

% change in $1/day poverty gap squared poverty	Full sample		Spell starting before 1995		Spell starting in 1995 or thereafter	
	Coeff.	p-value	Coeff.	p-value	Coeff.	p-value
test (p-value)						
Agriculture=agriculture × (poverty line/average income) = 0		0.000		0.000		0.073
Industry=industry × (poverty line/average/income) = 0		0.005		0.000		0.017
Service=service × (poverty line/average income) = 0		0.221		0.293		0.493
Agriculture+0.3 × agriculture = industry+ 0.3 × industry				0.000		0.268
Agriculture+0.7 × agriculture = industry+ 0.7 × industry				0.231		0.71
Agriculture+0.3 × agriculture = service+ 0.3 × service				0.009		0.094
Agriculture+0.7 × agriculture = service +0.7 × service				0.717		0.085

APPENDIX 4

Welfare Effect of Productivity and Output Price Change

Consider the utility function $u(q,L)$, which represents the preferences of a rural household[1] defined over consumption and work effort, with q representing the vector of goods consumed and L, the vector of labor supplied to each activity (including the household's own production activities). For example, $L=L_f+L_{nf}$ is the total labor supplied by the household to farm (L_f) and nonfarm (L_{nf}) activities. The household is assumed to maximize its utility by choosing the optimal consumption level (q) and the optimal leisure/labor trade-off (L), given its budget constraint. The marginal utility of consumption is positive, and the marginal utility of labor is negative.

To derive the monetary value of changes in agricultural productivity or food prices, the indirect utility function of the household—following Ravallion (1990), Deaton (1997), Chen and Ravallion (2003), and Minten and Barrett (2006)—is defined as

$$v(p,w,A) = \max_{q,L}[u(q,L)|\pi(p,w,A)+wL = p.q], \qquad (A4.1)$$

where p is the vector of prices for goods q, w is the vector of wage rates, and $\pi(p,w,A)$ is the profit obtained from all household enterprises as given by

$$\pi(p,w,A) = \max_{L^F}[pQ - wL^F | Q = Af(L^F, Z|E)], \qquad (A4.2)$$

where L^F is the total labor input into the household's own production activities, and A reflects the overall productivity of underlying production technology $f(.)$. A higher level of A implies a higher level of output per unit of cultivated land, Z, or per unit of labor employed in agriculture L^F, given underlying agroecological conditions (E). Under standard assumptions, an increase in A boosts agricultural output for those who have land ($Z>0$) and adequate agroecological conditions ($E>0$) and who allocate labor to agriculture production (L^F).[2] Rural labor markets are assumed to be efficient, and household and hired labor are substitutes, such that the household can recursively solve its utility maximization problem. It first chooses the optimal labor input in its own production activity ($L^F=L_f+H$, where H is the amount of hired labor) to maximize its profit and subsequently chooses q and L to maximize its utility, given its profit and income from other activities. The household is a price and wage taker. Prices and wages are given for the household, and for ease of presentation[3] consumer and producer prices are assumed to be the same.

Taking the total differential of equation A4.1 and applying the envelope theorem (whereby the welfare effects of small changes in the parameters in the neighborhood of an optimum can be evaluated by treating the quantity choices as given), the household's welfare gain from an increase in total factor productivity A is given by

$$g \equiv \frac{dv}{\varphi\, dA} = [Q-q]\frac{dp}{dA} + [L-L^F]\frac{dw}{dA} + p\frac{\partial Q}{\partial A}, \quad (A4.3)$$

where φ is the marginal utility of income, $L-L^F = L_{nf}-H$ is the household's net engagement in the external labor market (that is, the amount of the household's labor outside its own production activities reduced by the amount of labor hired in), and $Q-q$ is the household's net demand for its own produced good (for example, food). Similarly, the household's welfare gain from an increase in the price of the good is given by

$$g \equiv \frac{dv}{\varphi\, dp} = [Q-q] + [L-L^F]\frac{dw}{dP} + p\frac{\partial Q}{\partial p}\frac{dA}{dP}. \quad (A4.4)$$

Notes

1. Subscripts are dropped for ease of notation.
2. Weak monotonicity and weak concavity for $f(.)$ and $f(0)=0$ for any argument (that is, no output without labor, land, or essential biophysical inputs such as rain or soil nutrients) are assumed. For simplicity, the cost of inputs has been omitted.
3. To relax this assumption, assume that $p_p = \gamma p_c$, where p_c is the consumer price, p_p is the producer price, and $0 \leq \gamma \leq 1$.

Country Case Study Papers

Ethiopia

Demeke, Mulat. 2005. "Agricultural Performance and Policies in Ethiopia." Mimeo, World Bank, Washington, DC.

Dercon, Stefan, and Luc Christiaensen. 2005a. "Consumption Risk, Technology Adoption, and Poverty Traps: Evidence from Ethiopia." Mimeo, World Bank, Washington, DC.

Dercon, Stefan, and Luc Christiaensen. 2005b. "The Impact on Poverty of Policies to Stimulate Modern Input Adoption: The Case of Fertilizer in Ethiopia." Mimeo, World Bank, Washington, DC.

Kenya

Jayne, Thom, Robert J. Myers, and James Nyoro. 2005. "Effects of Government Maize Marketing and Trade Policies on Maize Market Prices in Kenya." Mimeo, World Bank, Washington, DC.

Mistiaen, Johan. 2006. "Poor and Inefficient? Farmers, Scientists, and the Maize Yield Gap in Kenya." Mimeo, World Bank, Washington, DC.

Nyoro, James K., Milu Muyanga, and Isaac Komo. 2005. "Revisiting the Role of Agricultural Performance in Reducing Poverty in Kenya." Mimeo, World Bank, Washington, DC.

Mude, Andrew Gache. 2005. "Estimating the Welfare Impact of Maize Price Policy in Kenya." Mimeo, World Bank, Washington, DC.

Madagascar

Minten, Bart, ed., with Christopher Barrett, Claude Randrianarisoa, Zará Randriamiarana, and Tiaray Razafimanantena. 2006. *The Role of Agriculture in Poverty Alleviation Revisited: The Case of Madagascar.* Washington, DC: World Bank.

Minten, Bart, Zará Randriamiarana, and Tiaray Razafimanantena. 2006. "Economic Growth, Agricultural Performance and Poverty in Madagascar." In *The Role of Agriculture in Poverty Alleviation Revisited: The Case of Madagascar,* ed. Bart Minten. Washington, DC: World Bank.

Minten, Bart, and Tiaray Razafimanantena. 2006. "The Structure of the Rural Economy." In *The Role of Agriculture in Poverty Alleviation Revisited: The Case of Madagascar,* ed. Bart Minten. Washington, DC: World Bank.

Minten, Bart, and Christopher B. Barrett. 2006. "Agricultural Productivity and Poverty Reduction." In *The Role of Agriculture in Poverty Alleviation Revisited: The Case of Madagascar,* ed. Bart Minten. Washington, DC: World Bank.

Randrianarisoa, Claude, and Bart Minten. 2006. "Improving Agricultural Productivity." In *The Role of Agriculture in Poverty Alleviation Revisited: The Case of Madagascar,* ed. Bart Minten. Washington, DC: World Bank.

Randrianarisoa, Claude, Tiaray Razafimanantena, and Bart Minten. 2006. "Simulating the Welfare Impact of Agriculture and Nonagriculture Investments." In *The Role of Agriculture in Poverty Alleviation Revisited: The Case of Madagascar,* ed. Bart Minten. Washington, DC: World Bank.

Minten, Bart. 2006. "How to Alleviate Rural Poverty through Agricultural Growth?" In *The Role of Agriculture in Poverty Alleviation Revisited: The Case of Madagascar,* ed. Bart Minten. Washington, DC: World Bank.

Tanzania

Sarris, Alexander, Sara Savastano, and Luc Christiaensen. 2006. "Agriculture and Poverty in Commodity-Dependent African Countries: A Household Perspective from Rural Tanzania." Commodities and Trade Technical Paper 9, Commodities and Trade Division, Food and Agriculture Organization, Rome.

Review paper

Christiaensen, Luc, Lionel Demery, and Jesper Kuhl. 2006. "The Role of Agriculture in Poverty Reduction: An Empirical Perspective." Policy Research Paper 4013, World Bank, Washington, DC.

References

Abrar, S., O. Morrissey, and T. Rayner. 2004. "Crop-Level Supply Response by Agro-Climatic Region in Ethiopia." *Journal of Agricultural Economics* 55 (2): 289–311.

Adams, Richard. 2004. "Economic Growth, Inequality, and Poverty: Estimating the Growth Elasticity of Poverty." *World Development* 32 (12): 1989–2004.

Ajibefun, Igbekele A., George Battese, and Adebiyi Daramola. 2002. "Determinants of Technical Efficiency in Smallholder Food Crop Farming: Application of Stochastic Frontier Production Function." *Quarterly Journal of International Agriculture* 41 (3): 225–40.

Ali, M., and D. Byerlee. 1991. "Economic Efficiency of Small Farmers in a Changing World: A Survey of Recent Evidence." *Journal of International Development* 3 (1): 1–27.

Anand, Sudhir, and S. M. R. Kanbur. 1985. "Poverty under the Kuznets Process." *Economic Journal* 95: 42–50.

Binam, Joachim, Jean Tonyè, Njankoua Wandji, Gwendoline Nyambi, and Mireille Akoa. 2004. "Factors Affecting the Technical Efficiency among Smallholder Farmers in the Slash and Burn Agriculture Zone of Cameroon." *Food Policy* 29 (5): 531–45.

Bourguignon, François, and Christian Morrisson. 1998. "Inequality and Development: The Role Dualism." *Journal of Development Economics* 57: 233–57.

Bourguignon, François. 2003. "The Growth Elasticity of Poverty Reduction: Explaining Heterogeneity across Countries and Time Periods." In *Growth and Inequality*, ed. T. Eichler and S. Turnovsky. Cambridge: MIT Press.

Boyce, J., and M. Ravallion. 1988. "Wage Determination in Rural Bangladesh: Effects of Relative Prices and Agricultural Productivity." Mimeo, Agricultural Policies Division, Policy, Planning and Research, World Bank, Washington, DC.

Bravo-Ortega, Claudio, and Daniel Lederman. 2005. "Agriculture and National Welfare around the World: Causality and International Heterogeneity since 1960." Policy Research Working Paper 3499, World Bank, Washington, DC.

Byerlee, Derek, Xinshen Diao, and Chris Jackson. 2005. "Agriculture, Rural Development, and Pro-Poor Growth." Discussion Paper 21, Agriculture and Rural Development, World Bank, Washington, DC.

Chen, S., and M. Ravallion. 2003. "Household Welfare Impacts of China's Accession to the World Trade Organization." Policy Research Working Paper 3040, World Bank, Washington, DC.

Christiaensen, Luc. 2004. "Costs and Benefits from Investing in Road Infrastructure? Case Evidence from Ethiopia." Mimeo, World Bank, Washington, DC.

Christiaensen, Luc, Lionel Demery, and Stefano Paternostro. 2005. "Reforms, Remoteness, and Risk in Africa: Understanding Inequality and Poverty during the 1990s." In *Spatial Inequality and Development*, ed. R. Kanbur and A. J. Venables, 209–36. UNU-WIDER Studies in Development Economics. Oxford: Oxford University Press.

Christiaensen, Luc, Vivian Hoffmann, and Alexander Sarris. 2005. "Coffee Price in Perspective: Household Vulnerability among Rural Coffee-Growing Smallholders in Rural Tanzania." Mimeo, World Bank, Washington, DC.

de Janvry, Alain, and Elisabeth Sadoulet. 2002. "World Poverty and the Role of Agricultural Technology: Direct and Indirect Effects." *Journal of Development Studies* 38 (4): 1–26.

Deaton, Angus, 1997. "The Analysis of Household Surveys: A Microeconometric Approach to Development Policy." Baltimore: World Bank and John Hopkins University Press.

Delgado, Christopher L., Jane Hopkins, and Valerie A. Kelly. 1998. "Agricultural Growth Linkages in Sub-Saharan Africa." Research Report 107, International Food Policy Research Institute, Washington, DC.

Dercon, Stefan, Daniel Gilligan, John Hoddinott, and Tassew Woldehanna. 2006. "The Impacts of Roads and Agricultural Extension on Crop Income, Consumption, and Poverty in Fifteen Ethiopian Villages." Mimeo, Oxford University, Oxford.

Dorosh, Paul, and Steven Haggblade. 2003. "Growth Linkages: Price Effects and Income Distribution in Sub-Saharan Africa." *Journal of African Economies* 12 (2): 207–35.

Ehui, Simon, and John Pender. 2005. "Resource Degradation, Low Agricultural Productivity, and Poverty in Sub-Saharan Africa: Pathways Out of the Spiral." *Agricultural Economics* 31 (2): 217–33.

Fan, S., C. Chan-Kang, and A. Mukherjee. 2005. "Rural and Urban Dynamics and Poverty: Evidence from China and India." Mimeo, International Food Policy Research Institute, Washington, DC.

Gollin, Douglas, Stephen Parente, and Richard Rogerson. 2002. "The Role of Agriculture in Development." *American Economic Review* 92 (May): 160–64.

Haggblade, Steven, Jeffrey Hammer, and Peter Hazell. 1991. "Modeling Agricultural Growth Multipliers." *American Journal of Agricultural Economics* 73 (2): 361–74.

Hayami, Yujiro, and Vernon Ruttan. 1985. *Agricultural Development: An International Perspective.* Baltimore: Johns Hopkins University Press.

InterAcademy Council. 2004. *Realizing the Promise and Potential of African Agriculture: Science and Technology Strategies for Improving Agricultural Productivity and Food Security in Africa.* Amsterdam: InterAcademy Council.

Irz, Xavier, and Terry Roe. 2005. "Seeds of Growth? Agricultural Productivity and the Transitional Dynamics of the Ramsey Model." *European Review of Agricultural Economics* 32 (2): 143–65.

Johnston, Bruce, and John Mellor. 1961. "The Role of Agriculture in Economic Development." *American Economic Review* 4: 566–93.

Jorgenson, Dale W., Frank M. Gollop, and Barbara M. Fraumeni. 1987. *Productivity and U.S. Economic Growth.* Cambridge, Massachusetts: Harvard University Press.

Kaliba, Aloyce. 2004. "Technical Efficiency of Smallholder Dairy Farms in Central Tanzania." *Quarterly Journal of International Agriculture* 43 (1): 39–55.

King, Robert P., and Derek Byerlee. 1978. "Factor Intensities and Locational Linkages of Rural Consumption Patterns in Sierra Leone." *American Journal of Agricultural Economics* 60 (2): 197–206.

Klasen, Stephan, and Mark Misselhorn. 2006. "Determinants of the Growth Semi-Elasticity of Poverty Reduction." Mimeo, University of Gottingen, Gottingen.

Krueger, Ann, Maurice Schiff, and Alberto Valdes. 1988. "Agricultural Incentives in Developing Countries: Measuring the Effect of Sectoral and Economywide Policies." *World Bank Economic Review* 2 (3): 255–71.

Lewis, P., W. Martin, and C. Savage. 1988. "Capital and Investment in the Agricultural Economy." *Quarterly Review of the Rural Economy* 10(1): 48–53.

Lipton, M. 1977. *Why Poor People Stay Poor: A Study of Urban Bias in World Development.* Canberra: Australian National University Press.

Lipton, Michael. 2001. "Rural Poverty Reduction: The Neglected Priority." Paper prepared for report on rural poverty by the International Fund for Agricultural Development, Rome.

Loayza, Norman, and Claudio Raddatz. 2005. "The Composition of Growth Matters for Poverty Alleviation." Mimeo, World Bank, Washington, DC.

Lopez, Ramon, and Alberto Valdes, eds. 2000. *Rural Poverty in Latin America.* New York: St. Martin's Press.

Martin, Will, and Devashish Mitra. 2001. "Productivity Growth and Convergence in Agriculture versus Manufacturing." *Economic Development and Cultural Change* 49 (2): 402–22.

Maxwell, S., I. Urey, and C. Ashley. 2001. "Emerging Issues in Rural Development: An Issues Paper." *Development Policy Review* 19 (4): 395–426.

Mellor, J. W. 1976. *The New Economics of Growth.* Ithaca, NY: Cornell University Press.

North, Douglas. 1959. "Agriculture in Regional Economic Growth." *Journal of Farm Economics* 41 (5): 943–51.

Perry, G., D. de Ferranti, David Lederman, William Foster, and Alberto Valdes. 2005. *Beyond the City: The Rural Contribution to Development—Complete Report.* Washington, DC: World Bank.

Ravallion, Martin. 1990. "Rural Welfare Effects of Food Price Changes under Induced Wage Responses: Theory and Evidence." *Oxford Economic Papers* 42 (3): 574–85.

———. 1997. "Can High-Inequality Countries Escape Poverty?" *Economic Letters* 56: 51–57.

———. 2001. "Growth, Inequality, and Poverty: Looking beyond Averages." *World Development* 29 (11): 1803–15.

———. 2003. "Measuring Aggregate Welfare in Developing Countries: How Well Do National Accounts and Surveys Agree?" *Review of Economics and Statistics* 85 (3): 645–52.

Ravallion, Martin, and Shaohua Chen. 2004. "How Have the World's Poorest Fared since the Early 1980s?" Policy Research Working Paper 3341, World Bank, Washington, DC.

———. 2007. "China's (Uneven) Progress against Poverty." *Journal of Development Economics.*

Ravallion, M., and G. Datt. 1996. "How Important to India's Poor Is the Sectoral Composition of Economic Growth?" *World Bank Economic Review* 10 (1): 1–25.

———. 1999. "When Is Growth Pro-Poor? Evidence from the Diverse Experiences of India's States." Policy Research Working Paper 2263, World Bank, Washington, DC.

———. 2002. "Why Has Economic Growth Been More Pro-Poor in Some States of India Than Others?" *Journal of Development Economics* 65: 381–400.

Ravallion, Martin, and Monica Huppi. 1991. "Measuring Changes in Poverty: A Methodological Case Study of Indonesia during an Adjustment Period." *World Bank Economic Review* 5 (January): 57–82.

Sarris, Alexander, Panayotis Karfakis, and Luc Christiaensen. 2006. "The Stated Benefits from Commodity Price and Weather-Based Insurance." Mimeo, World Bank, Washington, DC.

Schultz, Theodore W. 1964. *Transforming Traditional Agriculture*. New Haven, Connecticut: Yale University Press.

Sherlund, Shane, Christopher B. Barrett, and Adesina Akinwumi. 2002. "Smallholder Technical Efficiency Controlling for Environmental Production Conditions." *Journal of Development Economics* 69 (1): 85–101.

Thirtle, C., L. Irz, L. Lin, V. McKenzie-Hill, and S. Wiggins. 2001. "Relationship between Changes in Agricultural Productivity and the Incidence of Poverty in Developing Countries." Report commissioned by the U.K. Department for International Development, London.

Tiffen, Richard, and Xavier Irz. 2006. "Is Agriculture the Engine of Growth?" *Agricultural Economics* 35: 79–89.

Timmer, Peter. 1997. "How Well Do the Poor Connect to the Growth Process?" Discussion Paper 178, Consulting Assistance on Economic Reform project, Harvard Institute for International Development, Cambridge, Massachusetts.

———. 2005. "Agriculture and Pro-Poor Growth: What the Literature Says." Background paper prepared for the Operationalizing Pro-Poor Growth research program, World Bank, Washington, DC.

Yanggen, D., V. Kelly, T. Reardon, and A. Naseem. 1998. "Incentives for Fertilizer Use in Sub-Saharan Africa: A Review of Empirical Evidence on Fertilizer Yield Response and Profitability." International Development Working Paper 70, Michigan State University, East Lansing, Michigan.

Weber, M. T., John M. Staatz, John S. Holtzman, Eric W. Crawford, and Richard H. Bernsten. 1988. "Informing Food Security Decisions in Africa: Empirical Analysis and Policy Dialogue." *American Journal of Agricultural Economics* 70 (December): 1044–52.

Wood, Adrian. 2002. "Could Africa Be Like America?" Address to the Advisory Board of the Research Program on Enterprise Development, World Bank, Washington, DC.

World Bank. 2000. *The Quality of Growth*. Oxford: Oxford University Press.

World Bank. 2005a. World Development Indicators and Global Development Finance Database. World Bank, Washington DC.

World Bank. 2005b. "Ethiopia: Well-Being and Poverty in Ethiopia—The Role of Agriculture and Agency." Report 29468-ET, Africa Region, World Bank.

World Bank. 2005c. PovcalNet. http://iresearch.worldbank.org/PovcalNet/jsp/index.jsp.

World Bank. 2005d. "Tanzania: Sharing and Sustaining Economic Growth." Country Economic Memorandum and Poverty Assessment, World Bank, Washington, DC.

World Bank. 2006a. "Ethiopia: Rural Development Strategy." Mimeo, World Bank, Washington, DC.

World Bank. 2006b. "Promoting Increased Fertilizer Use in Africa: Lessons Learned and Good Practice Guidelines." Mimeo, Africa Region, World Bank, Washington, DC.

Index

Figures, notes, and tables are denoted by "f," "n," and "t" following page numbers.

A

access to credit, 59–61, 80
access to roads, xii, 62, 81
agricultural value added per worker and GDP growth per capita, 70
agriculture
  See also Sub-Saharan African agriculture; *specific countries*
  as backward subsistence sector, 1, 50
  differences between Asia and Sub-Saharan Africa, 4
  effect on nonagricultural sectors of economy, 67–71
  growth potential of, 47–65
    See also growth potential of agriculture
  investments in, effect of, 68–69, 73
  as livelihood of poor people, 4, 13, 79
  participation effect on poverty, 31–46
  role in poverty reduction. *See* role of agriculture in poverty reduction
agro-pessimism in Sub-Saharan Africa, 2, 4–5
allocative inefficiency affecting agricultural productivity, 52–53, 79
Asia
  See also specific countries and regions
  comparison with Africa. *See* Sub-Saharan African agriculture
  Green Revolution in, 1, 2, 3*f*
  poverty level in, 2, 3*f*
  productivity in nonagricultural sectors exceeding agriculture, 49

C

Cameroon, 65*n*
case studies
  See also Ethiopia; Kenya; Madagascar; Tanzania
  data sources and constructs used in, 85–88
  objectives of and data sources for, 83–84
Central Asia, 49, 76
cereal yields. *See specific countries and crops*
China, 4, 10, 14, 27, 45*n*

101

coffee growers, 60, 65n
conceptual framework, 9–12
contract farming and access to credit, 60, 61
costs and cost-benefit analysis, 80
 of fertilizer, 61
 of irrigation, 57–58
 of road construction, 62, 81
credit, access to, 59–61, 80
crop management techniques, improvement of, 53
crop value added per acre, effect on poverty reduction and consumption, 32–33, 45n, 46n
cross-country analysis
 growth potential of agriculture, 47–51, 63–64
 linking with cross-sector data, 15, 48t, 48–49, 49t, 69
 on poverty, xii, 5
 on sectoral growth, xii, 5, 14–15
cross-sector analysis
 GDP elasticities of poverty and, 14
 global comparison of nonagricultural vs. agricultural growth rates, 47–48, 48t, 74
 lack of data, effect of, 15
 linking with cross-country analysis, 15, 48t, 48–49, 49t
 population reallocation and poverty, 12n
 on poverty-reducing effect of growth, 16–31, 78
 share and elasticity components of participation effect on headcount poverty, 27, 28–29t, 30

D

data sources. *See* case studies
demand-side constraints on adoption of modern inputs, 59, 61

E

Eastern Europe, 49, 76
economic theory on agriculture, 1
education's effect
 on adoption of modern inputs, 59, 61
 on income inequality, 15
empirical considerations, 13–15, 85
Engel's Law, 51, 73, 78
Ethiopia
 access to credit in, 60
 agriculture's role in poverty reduction in, xii, 5, 31–32, 33
 cereal yields in, 52, 53, 59
 data sources and constructs used in case studies of, 87–88
 extension agents' visits in, 61
 fertilizer use in, 56t, 56–57, 58, 59, 60–61
 irrigation techniques in, 57
 location-specific approach to boosting cereal production in, 54
 objectives of and data sources for case studies of, 83
 road construction in, 62
 technical inefficiency in, 52, 53
 tradability of crops in, 42
 weather-related events and agricultural productivity in, 55, 80
experimental plots, yields of, 53
extension agents' visits, effect of, 61

F

fertilizer
 effect on agricultural productivity, 32, 33, 46n, 52, 56t, 56–57, 59, 80
 low adoption level of chemical fertilizer, 58, 61, 65n
 poorer farmers and, 63
 supply chain problems raising price for, 59
final demand effects, 11, 67
Food and Agricultural Organization of the United Nations, 17
food prices
 fertilizer use related to, 56
 as first-order effects, 37–39
 policies leading to increases and effect on poverty, 42–44, 43t, 46n
 as second-order effects, 39–42, 44
 wage rates, effect of changes on, 38–39, 79

G

GDP elasticities of poverty across sectors, 14, 15, 16–31, 45n, 77
globalization's effect on reducing advantage of agriculture over other sectors in reducing poverty, 27, 74, 78
government intervention on food prices, 38–39, 42–44, 43t, 46n, 79
gravitation irrigation, 57

Green Revolution in Asia, 1, 2
growth elasticity of poverty, 12n
growth potential of agriculture, 47–65
   See also participation effect
   cross-country perspective, 47–51
   effect of enhancing agricultural productivity, 51–64, 79–80
      access to roads, xii, 62, 81
      differences between richer and poorer farmers, 62–64
      efficiency gaps, 51–53
      need for location-specific analyses, 53–55, 80
      weather-related events, 55–58, 80

H

higher-income countries and participation effect of agriculture, 30, 74, 78
household welfare gain from enhanced food productivity, 36–39, 44, 91–92
human capital's distribution, effect on income inequality, 15

I

India, 9, 14, 15, 27, 36, 45n
indirect growth effects of agriculture, 4, 70, 73, 78
industry sector or industry and services sectors. *See* nonagricultural growth
inefficiency. *See* technical inefficiency affecting agricultural productivity
insurance for weather-based events, 57–58, 63, 80
intersectoral links, 11, 67
   production vs. consumption linkages, 68, 71n
investments in agriculture, effect of, 68–69, 73
irrigation. *See* water management and increased agricultural productivity

K

Kenya
   agriculture's role in poverty reduction in, xii, 5, 33
   cereal yields in, 52
   data sources and constructs used in case studies of, 87
   government intervention to stabilize food prices in, 42–44, 43t, 79
   objectives of and data sources for case studies of, 83–84
   second-order effects in, 31
   tradability of crops in, 42
Kuznets process, 12n

L

labor- vs. land-saving technology, variations across regions, 53–54
land distribution's effect on income inequality, 15, 54
Latin America, compared with African agricultural productivity, 68, 70
location-specific factors affecting agricultural productivity, 53–55, 80

M

Madagascar
   access to roads and adoption of technology in, 62
   agriculture's role in poverty reduction in, xii, 5
   cereal yields in, 52, 53
   data sources and constructs used in case studies of, 85–86
   economywide effects of investments in, 70
   fertilizer use in, 57, 58, 61, 80
   first-order effects in, 37–39
   irrigation in, 57, 61
   labor- vs. land-saving technology in, 54
   objectives of and data sources for case studies of, 84
   rice yields in, 39–42, 52, 53, 54, 58
   richer vs. poorer farmers in, 62–63
   second-order effects in, 31, 39–42
   spending on research to improve crop productivity, 53
   tradability of crops in, 42
   weather-related events and agricultural productivity in, 55, 80
maize yields
   in Ethiopia, 53, 59
   in Kenya, 52, 79
migration of labor out of agriculture, 77–78

N

net food buyers and sellers, 39, 40t, 44, 79
nonagricultural growth
   See also cross-sector analysis

compared with agricultural growth, 25, 30, 48–49, 48–49t
effect of agricultural growth on, 67–71
effect on agricultural growth of, 69
role in poverty reduction, 10, 11f, 14

P

participation effect
 agricultural growth's effect on, 31–44, 73, 77
  links to food prices and food productivity increases, 36–39, 44
  net food buyers and, 39, 40t, 44
  policies affecting food prices, 42–44, 43t
  price and wage effects, 39–42
 in conceptual framework, 9, 11–12
 theory and empirical literature on, 13–46
pest and crop disease management, effect of, 53, 59
poor people
 See also poverty data
 earning living from agriculture, 4, 13, 79
 labor as major asset of, 14
 located in rural areas, 13
 richer vs. poorer farmers, 62–64
population growth's contribution to productivity, 49
postharvest innovations, 53
poverty data
 by geographical coverage, 17, 17t
 headcount index, 17, 18, 19–21t, 45n, 74, 75t, 89–90t
 poverty gap index, 17, 18, 22–24t
 sectoral decomposition of poverty change, 73–76
 urban poverty compared to rural poverty, 70
Poverty Reduction Strategy Papers, xi, 2
price and wage effects of increased agricultural productivity, 39–42
 See also food prices

R

rice yields in Madagascar, 39–42, 52, 53, 54, 58
richer vs. poorer farmers, 62–64
road construction, xii, 62, 81

role of agriculture in poverty reduction
 importance of, xi, 4, 51, 77
 questioning of, 2, 48–49
 role of nonagricultural growth vs., 10, 11f, 14
 sectoral decomposition showing, 73–76, 75t
rural areas
 land and education in, 15
 likelihood of poor living in, 13
 neglect of infrastructure in, 4
 road construction in, xii, 62, 81
 urban poverty compared to rural poverty, 70

S

savings and credit organizations (SACCOs), 60, 61
second-order effects, 31, 33, 36, 44
services growth compared with agricultural growth, 25, 26–27, 45n
 See also nonagricultural growth
short-run elasticity of poverty, 36
Smith, Adam, 48
soil fertility innovations, 53, 64n
 See also fertilizer
SRI. *See* Système de Riz Intensive
Sub-Saharan African agriculture
 See also specific countries
 compared to Asia, 4, 15, 52, 68, 70
 effect on poverty reduction, 4, 26, 30–31, 63–64, 76, 77
 efficiency gap in, 51–53, 79
 fertilizer use in, 56–59
 focus points of study, 5
 globalization's effect on, 27
 income growth from agricultural sector in, 68
 limited productivity in, 2
 location-specific factors affecting productivity in, 53–55
 population growth's contribution to productivity in, 49
 productivity in, exceeding nonagricultural sectors in, 49
 yields in, 2, 3f, 51–53
Sub-Saharan African poverty, 2, 3f
 See also specific countries
 cross-sector analysis for poverty reduction, 26–31

link to agricultural growth, 4, 26, 30–31, 63–64, 76, 77
net food buyers and, 39, 40*t*, 44, 79
supply-side constraints on adoption of modern inputs, 59, 61
Système de Riz Intensive (SRI) cultivation technique, 52, 61, 64*n*

T

Tanzania
  access to credit in, 59
  agriculture's role in poverty reduction in, xii, 5, 32–33
  cereal yields in, 52, 53
  data sources and constructs used in case studies of, 86–87
  fertilizer use in, 58, 65*n*, 80
  irrigation techniques in, 57, 80
  land use variances in, 54
  low level of adoption of modern inputs in, 59–60
  objectives of and data sources for case studies of, 84
  rainfall's effect on agricultural productivity in, 55, 80
  tobacco farmers in, 60
technical inefficiency affecting agricultural productivity, 51, 52, 54, 64–65*n*, 79
theory and empirical considerations, 13–15, 85
tobacco farmers, 60

total factor productivity (TFP) growth in agriculture, 50
tradability of crops, 39–42

U

urban poverty compared to rural poverty, 70

V

Vietnam, 52

W

wage-goods effects, 11, 67, 70*n*
wage rates, effect of changes in food prices on, 38–39, 79
water management and increased agricultural productivity, xii, 55–58, 61, 63, 80
  effect on access to credit, 60
weather-related events and agricultural productivity, 55–58, 80
welfare effect of productivity. *See* household welfare gain from enhanced food productivity
World Bank
  Povcal database, 17
  World Development Indicators and Global Development Finance database, 17

ECO-AUDIT
Environmental Benefits Statement

The World Bank is committed to preserving endangered forests and natural resources. The Office of the Publisher has chosen to print *Down to Earth* on recycled paper with 30 percent postconsumer fiber in accordance with the recommended standards for paper usage set by the Green Press Initiative, a nonprofit program supporting publishers in using fiber that is not sourced from endangered forests. For more information, visit www.greenpressinitiative.org.

Saved:
- 11 trees
- 8 million BTUs of total energy
- 1,002 lbs. of net greenhouse gases
- 4,161 gallons of wastewater
- 534 lbs. of solid waste

green press INITIATIVE